ENCYCLOPEDIA OF GENOMICS

ENCYCLOPEDIA OF GENOMICS

Vol. 5

GENOMIC STRUCTURE

By

Dr. Shiv Kant Prasad

School of Studies of Zoology & Biotechnology

Vikram University

Ujjain

(India)

D P H

DISCOVERY PUBLISHING HOUSE PVT. LTD.

NEW DELHI-110 002

Published by:
Tilak Wasan
DISCOVERY PUBLISHING HOUSE PVT. LTD.
4383/4A, Ansari Road, Darya Ganj
New Delhi-110 002 (India)
Phone: +91-11-23279245, 43596064-65
Fax : +91-11-23253475
E-mail: parul.wasan@gmail.com
discoverypublishinghouse@gmail.com
web : www.discoverypublishinggroup.com

***First Edition:* 2012**

ISBN: 978-93-5056-032-7 (Set)
978-93-5056-033-4 (Vol. 1)
978-93-5056-034-1 (Vol. 2)
978-93-5056-035-8 (Vol. 3)
978-93-5056-036-5 (Vol. 4)
978-93-5056-037-2 (Vol. 5)

Encyclopedia of Genomics

Printed at:
Shree Balaji Art Press
Delhi

Preface

Progress in any scientific discipline is dependent on the availability of techniques and methods that extend the range and sophistication of experiments which may be performed. Over the last 30 years or so this has been demonstrated in a spectacular way by the emergence of genetic engineering. This field has grown rapidly to the point where, in many laboratories around the world, it is now routine practice to isolate a specific DNA fragment from the genome of an organism, determine its base sequence, and assess its function. The technology is also now used in many other applications, including pharmaceutical analysis of samples, paternity disputes, medical diagnosis, genomic and the biotechnology industry.

The present title **"Encyclopedia of Genomics"** is a fast growing area of research for undergraduate and post-graduate students in life sciences, chemistry, engineering and medicine. The overall objective of this publication is to provide a professional level reference work with comprehensive coverage of the molecular basis of genome and the application of that knowledge in genetics, evolution, medicine, and agriculture. It deals with the life processes at a molecular level genetic disease diagnosis and genetic therapy; the theory and techniques for understanding manipulating, and synthesizing biological molecules and their aggregates; and the application of biological processes to make or modify products to improve plants or animals or to develop microorganisms for specific uses.

The development of techniques for genomic science is based on an understanding of the molecular aspect of genomic processes occurring in nature. These techniques have, in turn, enabled expansion of our knowledge and understanding of how genes are organized and expressed in living cells and has provided major impetus to the development of biotechnological enterprises.

The present title has been written to provide readers with a thorough understanding of the techniques which underpin the developments in the field of genomic science. By gaining this understanding, readers will be able to apply the techniques of genomic manipulation to tackle specific problems or to achieve desired objectives in a wide variety of fields.

There can be no claim to originality except in the manner or treatment and much of the information has been obtained from the books and scientific journals available in the different libraries.

The author express his thanks to his friends and colleagues whose continue inspirations have initiated him to bring out this book.

The author expresses her gratitude to Mr. Wasan and staff of M/s Discovery Publishing House Pvt. Ltd. for their whole hearted co-operation in the publication of this book.

S.K. Prasad

Contents

Chapter 1

Aberration of Unusual Genome

Depending on conditions suceh as *transcription,* supercoil stress, protein binding, and so on. Some of the DNA sequences in the *genome* may exit ether in the *canonical* right-handed B duplex or in *alternative* non-B *conformations Analyses* of breakpoint junctions at deletions, *translocations,* and inversions, where the sites of DNA breakage could be determined at the nucleotide level, revealed that most, if not all, of the breaks occurred within, or adjacent to, the predicted non-B conformations. These *findings* support a model whereby rearrangements arc caused by recombination/repair processes between two distinct non-B conformations, which may reside either on the same chromosome or on two distinct chromosomes.

This model was applicable to both *Escherichia coli* and humans, suggesting that the mechanisms involved are highly conserved. The number of genomic *disorders* that are *explicable* in terms of non-B DNA conformation-mediated rearrangement is considerable (approx 30). Most involve *recombination* within (or between) *chromosomal* regions enriched in low-copy repeats (LCRs), whereas l*ymphoid-specific* rearrangements depend on the *recombination-activated* gene *(RAG)* recombinase.

However, considering the diseases associated with *triplet* repeat expansions, which are also believed to be mediated by non-B conformations, the total number of *pathological* conditions is around 50. This number may increase as further investigations are conducted on additional *rearrangements.* The history of investigations on DNA conformations that differ from the *canonical* right-handed B-form, as related to genetic diseases, dates back to the mid- 1960s.

Early studies with high-molecular-weight DNA polymers of defined repeating nucleotide sequences demonstrated the role of DN A sequence in their properties and *conformations.* Investigations with repeating homo-, di-, tri-, and *tetranucleotide* motifs revealed the *important* role of DNA sequence in molecular behaviours.

At that time, this concept was heretical because numerous prior physical and *biochemical investigations* with naturally occurring DNA sequences did not enable the realization of the effect of sequence. It should be noted that these *biochemical/ biophysical* studies in the 1960s predated DNA sequencing by at least a decade. Early studies were followed by a number of innovative

Name	Conformation	General Seq.. Requirements	Sequence
Cruciform	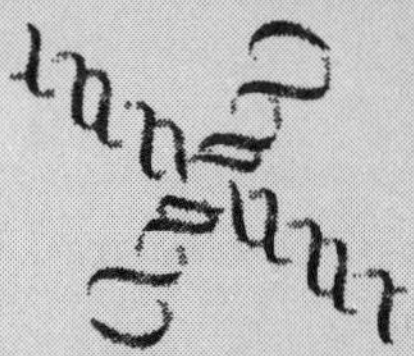	Inverted Repeats	TCGGTACCGA AGCCATGGCT
Triplex		$(R•Y)_n$ Mirror Repeats	AAGAGG ¦ GGAGAA TTCTCC ¦ CCTCTT
Sliped (Hairpin) Structure	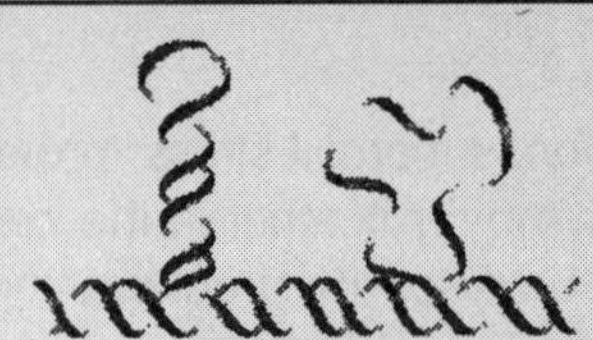	Direct Repeats	TCGGTTCGGT AGCCAAGCCA
Tetraplex	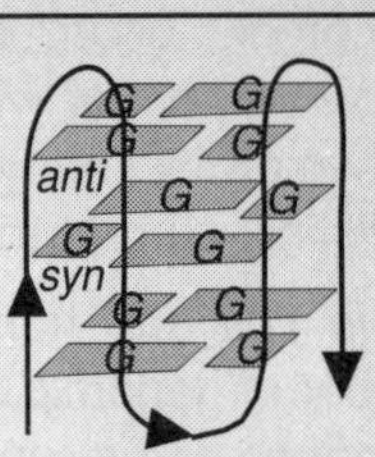	Oligo $(G)_n$ Tracts	$AG_3(T_2AG_3)_3$ single strand
Left handed Z - DNA	B - Z Junctions 	$(YR•YR)_n$	CGCGTGCGTGTG GCGCACGCACAC

Figure 1.1: Rearrangement of Unusual DNA.

discoveries on DNA conformational features in synthetic *oligomers, restriction fragments,* and *recombinant* DNAs.

The DNA *polymorphisms* were shown to be a function of sequence, *topology* (*supercoil density*), *ionic conditions, protein binding, methylation, carcinogen binding,* and other factors. A number of non-B DNA structures have been discovered, *approximately* one new conformation every 3 years during the past 35 years, and include triplexes, left-handed DNA, bent DNA, *cruciforms,* nodule DNA, *flexible* and writhed DNA, G4 tetrad (*tetraplexes*), slipped structures, and sticky DNA. From the outset, it has been realized U,27 that these sequence effects probably have profound biological implications, and indeed their role in transcription and in the maintenance of telomere ends has been *reviewed* recently.

Moreover, in the past few years, *dramatic* advances in genomics, *human genetics, physiology,* and DNA structural biology have revealed the role of non-B conformations in the etiology of at least 46 human genetic diseases that involve genomic *rearrangements*, as well as other types of *mutation* events.

CONFORMATIONS OF UNUSUAL DNA

Segments of DNA are *polymorphic*. A large number of simple DNA repeat sequences can exist in at least two conformations that are *interconvertable*. All of these sequences adopt the *orthodox* right-handed B -form, probably for the majority of the time, with Watson-Crick (WC) A•T and G•C basepairs. However, perhaps transiently, at least 10 non-B conformations are formed at specific sequence motifs as a function of negative supercoil density, generated in part by *transcription*, protein binding, or other factors.

Non-B DNA structures include cruciforms and left-handed Z-DNA, which have by in the WC *conformation*, whereas *triplexes*, *tetraplexes*, and *slipped* structures rely in part, or exclusively, on non-WC *interactions*. Cruciform DNA occurs at inverted repeats (IR), which are defined as sequences of identical composition on the *complementary* strands. Each strand folds at the IR center of symmetry and reconstitutes an *intramolecular* B-helix capped by a single-stranded loop that may extend from a few *basepairs* to several kilobases.

The term cruciform originates from the appearance of the duplex arms which at the four-way junction adopts either an "open" configuration that allows strand migration or a "stacked" (*locked*) arrangement, where the *helices* stack on each other. In both cases, the overall *conformation* and the intra-duplex angles mimic the Holliday junction, and in fact these structures have been studied extensively as models of *recombination intermediates*. *Cruciform* DNA is often incorrectly referred to as a *palindrome*; a *palindrome* is a word or sentence that reads identically when spelled backwards, such as rotor or *Hannah*.

Triplex (triple-helix, H-DNA) DNA occurs at oligo(purine•pyrimidine) [$(R•Y)_n$] tracts and is favoured by sequences containing a mirror repeat *symmetry*. The *purine* strand of the WC duplex engages the third strand through Hoogsteen hydrogen bonds in the major groove while maintaining the original duplex structure in a B-like *conformation*. Triplex DNA is *polymorphic* in the *orientation* and composition of the third strand; this may be either *pyrimidine*-rich and parallel to the *complementary* R-strand or *purine-rich* and antiparallel to the complementary R-strand.

In addition, the participating third strand may originate from within a single $(R•Y)_n$ tract (*intramolecular triplex*) or from a separate tract (*intermolecular triplex*). Direct repeats (DR) aligning out of register give rise to slipped structures with looped-out bases '). When DR involve several repeating motifs, like the telomeric or triplet repeat sequences, the looped-out bases may form duplexes stabilized by unusual pairs, such as T•T in $(CTG)_n$ or shared paired motifs (G•A and GC-AA), a specific arrangement characterized by interstrand, rather than *intrastrand, stacking interactions*.

A guanine tetrad, as found in telomeric DNA, is a square *coplanar* array of four guanines. Each base donates two *Hoogsteen hydrogen* bonds to one *neighbour* and receives two hydrogen bonds from the other neighbour in a cyclic arrangement involving N1, N2, O6, and N7.

Four-stranded DNA (tetraplex DNA, G-quadruplex) is composed of *guanine tetrads* stacked

Table 1.1: Syndrome, Sequence Alteration and DNA Repeat Motifs.

Sequence Alteration	Syndrome	DNA motifs
t(11;22)(q23;q11.2)	Supernumerary der(22) syndrome	IR
t(17;22)(q l 1.2;q] 1.2)	Neurofibromatosis type I	IR
t(1;22)(p21.2;q 11.2)	Ependymoma	IR
t(4;22)(835. 11 ;q 11.2)	Velocardiofacial syndrome	IR
del(22)(ql 1.2q11.2)	DiGeorge, velocardiofacial, or conotruncal anomaly face syndrome	IR, DR
inv dup(22) (q11.2q11.2)	Cat-eye syndrome	IR, DR
del(Yq)	Spermatogenic failure	IR, DR
i(17q)	Hematologic malignances (chronic myeloid leukemia)	IR
del(17)(p11.2p11.2)	Smith-Magenis syndrome	IR, DR
dup(17)(p11.2p11.2)	dup(ldup(1) syndrome	IR. DR
dup(17)(pl2p12)	Charcot-Marie-Tooth type 11	IR, DR
del(17)(p12p12)	Hereditary neuropathy with liability to pressure palsies	IR, DR
t(14;18)(q32.3;q21.3)	Follicular (B cells) lymphomas	(R-Y)-rich
del(7)(ql 1.23q11.23)	Williams-Beuren syndrome	IR, DR
inv(7)(ql1.23q11.23)	Predisposition to Williams-Beuren syndrome	IR
del(J5)(q11q13)	Prader-Willi and Angelman syndromes	IR, DR
t(X;22)(q27;q11)	Myeloschizis and lumbosacral spina bifida	IR
Expanded (CTG•CAG)n (coding)	Expanded polyglutamine diseases (Haw River, Huntington, Huntington's disease-like 2, spinobulbar muscular atrophy [Kennedy], spino-cerebellar ataxia [SCA] 1, SCA2, SCA3, SCA6, SCAT, SCA17)	DR
Expanded (C TG•CAG)n (non-coding)	Myotonic dystrophy type 1, SCA8, SCA12	DR
Expanded (CGG•CCG)n (coding)	Expanded polyalanine diseases (infantile spasm, cleidocranial dysplasia, blepharophimosis/ptosis/ epicanthus inversus type B, hand-foot-genital, synpolydactyly, oculopharyngeal muscular dystrophy, holoprosencephaly, oculopharyngeal muscular dystrophy)	DR
Expanded (CGG•CCG)n	Fragile XA, fragile XE. fragile XF, Jacobsen (FRA11 11B) DR	

(Table Contd.)

Sequence Alteration	Syndrome	DNA motifs
(non-coding)		
Expanded (GAA•TTC)n	Friedreich's ataxia	DR with (R•Y)
Expanded - (GAC•GTC)n	Pseudoachondroplasia, multiple epiphyseal dysplasia	DR
Expanded (CCTG•CAGG)n	Myotonic dystrophy type 2	DR
Expanded (ATTCT•AGAAT)n	SCA 10	DR
Expanded (CCCCGCCCCGCG)n	Progressive myoclonus epilepsy type I	DR
Expanded 24mer	Creutzfeldt-Jacob disease	DR
Contracted 3.3-kb D4Z4 repeat	Facioscapulohumeral muscular dystrophy	DR

on each other, and, therefore, is favoured by four runs of three or more guanines. This conformation is also endowed with a high degree of *polymorphism* in terms of strand *stoichiometry* and *polarity, glycosidic torsion angles, groove size*, and *connecting loops*.

Pairing schemes with other bases have also been observed, such as a *guanine tetrad sandwiched* between two G•C and A•T WC pairs. Left-handed Z-DNA is adopted by alternating pyrimidine-purine [$(YR•YR)_n$] sequences, such as $(CG•CG)_n$ and $(CA•TG)_n$.

The transition from the right-handed B- to the left-handed Z-form is accomplished by a 180° flip (upside-down) of the *basepair* through the rotation of every other *purine* from the *anti* to the svn *conformation* and a corresponding change in the sugar *puckering* mode from the C2'- to the C3'-*endo* conformation.

Contrary to the B-DNA, which possesses a major and a minor groove, Z-DNA contains only one groove. Other architectures (sticky DNA, nodule DNA, bent DNA, i-motifs, and others well known from a *biochemical standpoint*; their biological roles in *genome instability* are under *current investigation*.

STRUCTURAL TRANSLOCATION

Non-B DNA Conformations Cause Rearrangements

The involvement of double-strand breaks (DSB) in genome instability is well-documented, but the initiating events that lead to their formation are not fully *understood*. Recently, an increasing amount of information has revealed a major causative role for non-B DNA conformations. Indeed, *analyses* of large *deletions stimulated* by a 2.5 kb poly(R•Y) sequence from intron 21 of the polycystic *kidney* disease 1 gene *(PKD1)*, or by long (CTG•CAG), repeats, in *E. coli* indicated that the breakpoints occurred invariably at nt abutting, or within, *motifs* capable of adopting non-B conformations.

Because the same *PKD1* tract also induced cell death in a supercoil-dependent manner (18), we concluded that the large deletions were *mediated* by supercoil-dependent non-B conformations. In addition, the presence of short (direct or inverted) homologies at the breakpoint *junctions* suggested a model whereby two distantly located non-B DNA conformations induced DSB repair and, hence, demarcated the deletion endpoints.

Similarly, a search of 222 rearrangement breakpoints from the human Gross Rearrangement Breakpoint Database (GRaBD) (17) revealed that both the frequency and the proximity of (R•Y)n and (RY•RY)n tracts to breakpoint junctions were significantly greater than expected by chance.

A detailed analysis of the DNA sequences *flanking* the breakpoints in 11 deletions showed that, as in the bacterial cases, the events were *explicable* in terms of non-B DNA formation. Hence, non-B conformations appear to play a widespread role in rearrangement *mutagenesis.*

Chromosomal Rearrangements

A second line of evidence in establishing the role for DNA structural features in rearrangements was the recent identification of extremely large blocks of repeated sequences termed LCRs (*duplicons, segmental duplications*). Such blocks constitute a fertile substrate for recurrent rearrangements associated with more than 30 human *genomic* disorders because they may adopt conformations of *unprecedented complexity* and size.

Chromosome 22q11 is characterized by four large LCRs (LCR-A to D) that *encompass* approx 350 kb of DNA over an approx 3-Mb region. Sequence analyses of *constitutional translocation* breakpoints between chromosomes 22 and 1, 4, 11, or 17 have shown the *hallmarks* of cruciform DNA structures. For example, the breakpoints on LCR-B clustered in a narrow IR A/T-rich region that, in conjunction with flanking nuclear factor-1-like sequences, is *predicted* to form an approx 600 by-long *cruciform.*

Furthermore, the *clustering* of breakpoints occurred within 15 by from the IR center and involved the symmetric deletion of a few basepair on either side of the IR apex, consistent with the *prediction* that *cruciform loops* are susceptible to cleavage, and, hence, to repair. The breakpoints on chromosomes 1p21.2, 4q35. 1, 11q23, and 17q11.2, also occurred at, or near, the centers of IR. A/T-rich sequences.

Sperm analyses confirm that the t(11;22) translocation recurs at high frequency (approx 1-9 x 10^{-5}) in the cell population as *a de novo* mutation and, concurrently, in vitro studies provide direct evidence that the chromosome 11 IR has the *capacity* to fold into stable *cruciform conformations.*

These data support the strong *destabilizing* role for the sequences/structures involved. Interestingly, a case of *ependymoma* associated with the t(1;22) *translocation* revealed that the breakpoint on chromosome 1 was flanked by six copies of a 33 by-long motif with IR symmetry.

Conversely, the human genome assembly sequenced clone indicated the presence of only two such repeats; if the number of repeats is indeed *polymorphic* in the general population, this finding suggests that expansion may *predispose* to DNA instability, possibly by increasing the *propensity* (and/or stability) for *cruciform* formation.

Recombination events between highly *homologous* (97-98%) sequences within LCR-A to D are associated with recurrent deletions and duplications (approx 1:3000 live births), leading to the

DiGeorge, velocardiofacial, conotruncal anomaly face and cat-eye syndromes, suggesting that non-B DNA *conformations* may also be involved in these *conditions.*

Rearrangement and Anti Male Sterlity

Interstitial deletions in male-specific regions on chromosome Yq are a major source of *infertility* (AZF) Of the commonly affected regions (AZFa, AZFb and AZFc), AZFb and AZFc are located within an approx 8-Mb interval and are mostly *comprised* of five large IR (P1-P5), each consisting of a complex array of direct and inverted blocks *harboring male-specific* genes.

Large (>90 kb) *regions* of near-perfect identity exist among these IRs; *nevertheless,* the breakpoints strongly cluster toward the IR spacers (i.e., the regions that separate the two inverted repeat sequences) where, in some cases, they took place between sequences that shared no *homology*. These features are *consistent* with a cruciform-mediated *mechanism* for the deletion.

Rearrangements of Chromosomal Segments Causing Orchogenic Effect

Isochromosomes are abnormal chromosomes that contain a duplication of a single chromosome arm. Isochromosome 17q [i(17q)] is the most common *isochromosome* associated with human *neoplasia,* including *neuroectodermal tumors* and *medullablastomas,* representing approx 3% of single or additional chromosomal abnormalities in carcinomas and *hematologic malignancies,* and occur in approx 21% of such cases during *chronic myeloid leukemia* disease progression.

The i(17q) breakpoints cluster within 17p11.2, either in the very *pericentromeric* section or within the approx 4-Mb Smith-Magenis Syndrome common deletion region. An approx 240-kb interval (RNU3) encompassing the i(17q) breakpoint *cluster* region contains two LCRs, REPA and REPB, whereby two *inverted* copies of REPA (REPA 1 and REPA2, approx 38 kb) separated by a single copy of REPB (REPB3, approx 48 kb), are followed by two inverted copies of REPB (REPB2, approx 43 kb, and REPB 1, approx 49 kb). This arrangement provides the structural basis for the formation of large *cruciform* structures.

The i(17q) is consistent with a *recombination* mechanism between sister-chromatids mediated by a REPB1-REPB2 cruciform, which *predicts formation* of the dicentric i(17q) and an *acentric* chromosome that is expected to be lost P-21-The i(17q) RNU3 LCRs are part of a complex array of LCRs involved in deletions and duplications leading to the Smith-Magenis *Syndrome* the dup(17)(p11.2p11.2) syndrome, *hereditary neuropathy* with *liability* to pressure palsies (HNPP), and the Charcot-Marie-Tooth type 1 disease (CMT1A). Also in these cases, the presence of strand exchange *hotspots* strongly favours the concept of the formation of non-B DNA *conformations* as the molecular *rigger* for the rearrangements

Resequencing Causing Malignancy in Lymph Glands

During V(D)J *recombination,* V, D, and J coding segments are fused in varying *permutations* to create *antigen* receptor variable region exons with diverse antigen-binding potentials. The V, D, and J coding segments are flanked by *recombination* signal sequences (RSS), which consist of conserved *heptamer* and *nonamer motifs* separated by a *spacer* region of either 12 or 23 bp.

The RSS are specifically recognized by the RAG complex, which binds to two separated sequences to effect a DSB-dependent *recombination* reaction. *Nonauthentic* RSS may also be targeted for recombination, and specific chromosomal rearrangements involving Ig/TCR antigen receptor genes and proto-oncogenes, such as *LMO2, Ttg-1,* and *SIL* are caused by such a *promiscuous* RAG activity.

These chromosomal abnormalities are a major feature of *lymphoid neoplasms*. The t(14;18)(q32.3;q21.3) *translocation* is the most common translocation in cancer and accounts for almost all *follicular* lymphomas.

This chromosomal *aberration*, which is also owing to RAG activity, does not rely on *degenerate* RSS motifs. The gene involved on chr18 is the proto-oncogene *bcl-2*, which is generally fused to one of the J_H gene segments from the *immunoglobulin* heavy chain IgH locus on chr14 to *generate* the *derivative* chromosome 14.

Der(18) is generated by the joining of a D_H gene segment from the IgH locus on chr14 with the remaining 3' bcl-2 *untranslated* region exon on chrl8. Despite the size of the *bcl-2* gene (>200 kb), 75% of translocations occur within 150 by located in the untranslated portion of the *exon* three major breakpoint region (Mbr). Within these 150 bp, three peaks of *breakage* take place, each 15-20 by in size, whose sequence adopt a non-B DNA *conformation*.

The nature of the non-B *conformation*(s), which is under investigation, consists of *Hoogsteen hydrogen* bonding and one or more single-stranded regions that generate a structure *recognized* by the RAG complex.

Additional diseases believed to be mediated by non-B conformations are listed in Table elsewhere in this chapter. We should emphasize that, whereas rigorous proof exists for the structural *properties* of nonB DNA *conformations* as well as for the genome rearrangements and their involvement in human disease, direct *biophysical* determinations to *definitively demonstrate* the involvement of these non-B elements remain to be performed in most cases.

Obviously, the most *compelling evidence* would be derived from in vivo *conformational determinations*; however, these analyses are extremely difficult and hence have only been performed in limited cases. Nonetheless, even the more *facile* in vitro studies are *lacking* in most cases.

Neurological Disorders

At least 20 hereditary neurological diseases are caused by the expansion of simple triplet repeat sequences in either coding or non-coding regions. These expansions are mediated by DNA replication, repair, and *recombination*.

TDNA slipped structures play a prominent role in all of these mechanisms and the *preferential* formation of *hairpin* loops with long repeating CTG, GGC, and GAA strands, compared to their complementary strands, are critical to the *expansion* and deletion *behaviours* (*genetic instabilities*).

Furthermore, the flexible and writhed conformations adopted by long CTG•CAG and CCG•CGG tracts promote these behaviours. Friedreich's ataxia (FRDA) is the only *hereditary neurological* disease caused by an expanded GAA•TTC sequence. This tract, which occurs in intron 1, is well known to adopt the triplex *conformation* as well as the more *complex sticky* DNA structure between two long GAA•TTC tracts at distant *locations* of the same *molecule*.

All workers in this field agree that the triplex and/or sticky DNA conformation adopted by this repeating sequence is involved in the *inhibition* of transcription of the FRDA gene to give a *reduction* in the amount of the *frataxin protein*. Accordingly, the in vivo *discovery* of sticky DNA along with its molecular biological implications may be one of the more *convincing* cases of a non-B DNA structure involved in disease *pathology*.

CONCLUDING REMARKS

Dramatic advances were realized from the combined efforts of biophysics, human genetics, genomics, and molecular biological studies in our *comprehension* of the role of non-B DNA *conformations* in mutagenesis involved in genetic diseases. The *thesis* of this chapter is that, in general, all segments of DNA are usually in the orthodox right-handed B-conformation; however, certain repeat tracts *occasionally* adopt a non-B structure as a function of *supercoil density* or protein binding, and so on.

These non-B conformations (*cruciforms, triplexes, tetraplexes*, and others) are recognized by DNA repair systems and serve as breakpoints for several types of genome rearrangements, which are the genetic basis for a number of diseases. Future *emphases* will attempt to *fine-tune* the adoption of the non-B conformations at specific loci in order to correlate more rigorously this behaviour with the generation of *rearrangement endpoints*.

In addition, analyses may be performed to identify the types of non-B DNA structures that elicit certain kinds of *mutations* and the enzyme systems *involved*. Many more diseases *probably* will be *recognized* as owing to *mutations* at non-B DNA structures.

Specific DNA sequences adopt a number of alternative conformations (*cruciforms, triplexes, slipped structures, tetraplexes*, Z-DNA), also called non-B DNA, which determine the breakpoints responsible for *chromosomal* rearrangements and other *mutations* at sites of homology. The non-B DNA structures, rather than the sequences *per se*, are the likely *triggers* for these *recombination/ repair* functions. More thans 40 human *genetic* disorders may be *initiated* by these processes.

Chapter 2

Bacterial Genome Sequencing

Most bacterial genomes range from 0.5 to 7.0 Mb in size, and a random shotgun technique is the most effective sequencing strategy, except for organisms whose genomes contain multiple DNA repeats. Genome DNA, which may sometimes contain extrachromosomal DNA such as *plasmids,* is randomly sheared by ultrasonication or by passage under high pressure through a narrow slit, and subjected to agarose gel electrophoresis to yield short (1-2 kb), medium (4-10 kb), and long (approx 100 kb) fragments.

Short DNA fragments are useful for determining a draft sequence of the whole genome. The sheared genome DNA fragments may have *cohesive* (sticky) *ends,* and it is necessary to *"polish"* them with a *DNA exonuclease* (e.g., BAL31 nuclease) and the *Klenow fragment of DNA poly merase.* The resulting blunted DNA fragments are ready for insertion into a cloning vector (e.g., pUC 18).

Several host strains of *Escherichia coli* are available *commercially* for genomic library construction; the strain used should have a high *transformation* efficiency (>1 × 10^8 transformants/ μg sample DNA), and be able to accept methylated DNA.

Natural bacteria often have restriction-modification systems involving DNA digestion and *methylation* that protect their genome integrity from exogenous DNA. Strains of *E. coli,* such as DH5a*MCR or DH1OB, carry *mutations* in *mcrABC* and *mrr,* and are suitable hosts for cloning methylated DNA.

Once the fragments have been cloned, they are amplified from transformants in the genomic library by polymerise chain reaction (PCR), using vector-specific primers (e.g., M13 primers) and *purified* to act as templates for sequencing.

INDIVIDUAL SEQUENCES

The large number of reads is assembled with computer programs, such as the base-calling program phred and the assembler program phrap, into larger fragment sequences called *contigs.* This is achieved by making use of *overlapping* or common sequences within reads, and results in the production of an *integrated* map of the *chromosome.* At first the number of contigs increases

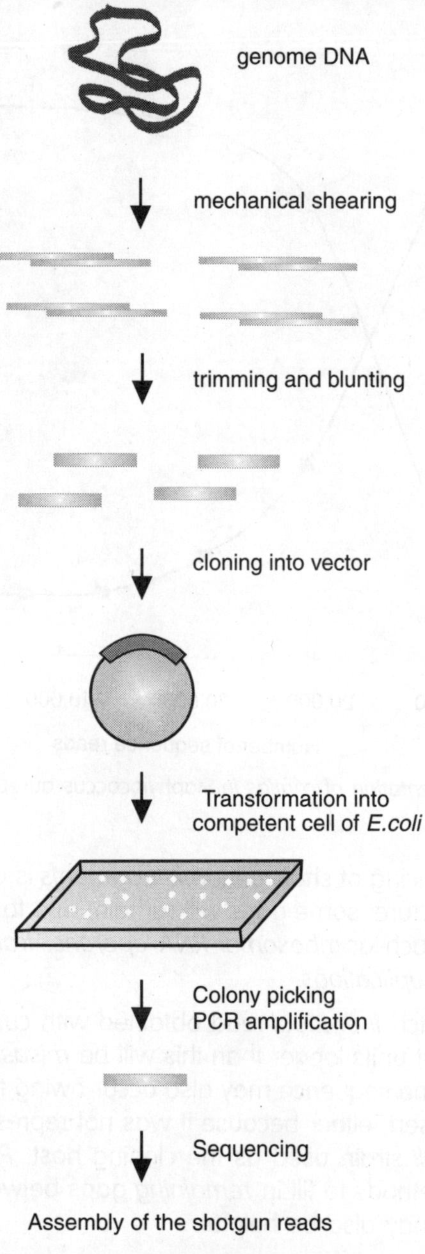

Figure 2.1: Showing the basic technique to calculate the DNA sequencing by short gun method.

during the assembly of reads, but overlapping sequences between *contigs* allows them to be further assembled, *resulting* in a subsequent *gradual* decrease in number.

During analysis of the *Staphylococcus aureus* genome, this decrease in the number of contigs was attenuated after 40,000 reads, *indicating* that any further sequencing of shotgun *clones* would not be expected to add new sequences.

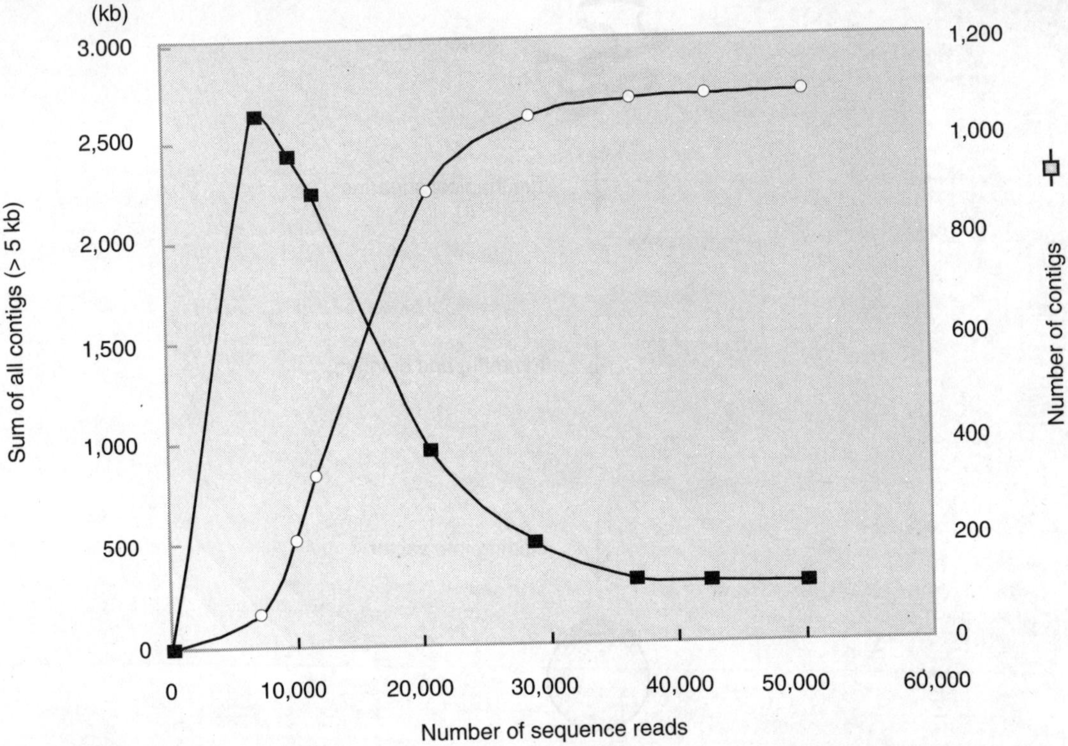

Figure 2.2: Graphical Presentation of contigs in Staphylococcus aureus *N315 genome sequencing.*

Generating Gap

Although shotgun sequencing of short *genomic fragments* is a fast method for determining a *crude* image of genome structure, some gaps will remain, due to the presence in the genome of *multiple-copy* repeat units, such as *ribosomal RNA operons, transposons, insertion sequences, bacteriophages,* and *gene duplications.*

The average length of each *individual* read obtained with current DNA *sequencing* methods is approx 500-700 bp. Repeat units longer than this will be *misassembled* into *one locus, leaving gaps.* Gaps in the true genome sequence may also occur owing to the lack of a clone *containing* a particular genomic DNA insert, either because it was not represented in the library or because of gene toxicity for the *E. coli* strain used as the cloning host. *Physical mapping* and long-PCR amplification are effective methods to fill in *remaining* gaps between contigs. Medium-length (4-10 kb) *shotgun* sequencing may also be helpful.

The multicopy repeat unit most commonly found in bacteria is the rRNA operon. The number of copies varies among species; for example, *S. caureus* possesses five or six copies, while mycobacteria and some other genera contain only a single copy. The *operon* is composed of genes *encoding* 5S, 16S, and 23S rRNA, and tRNAs. The detailed compositions of the rRNAs and tRNAs differ in each operon, and therefore sequencing templates containing individual rRNA operons should be obtained by *phage subcloning* or long-PCR amplification, to determine their exact structure and *localization* in the *whole-genome* sequence.

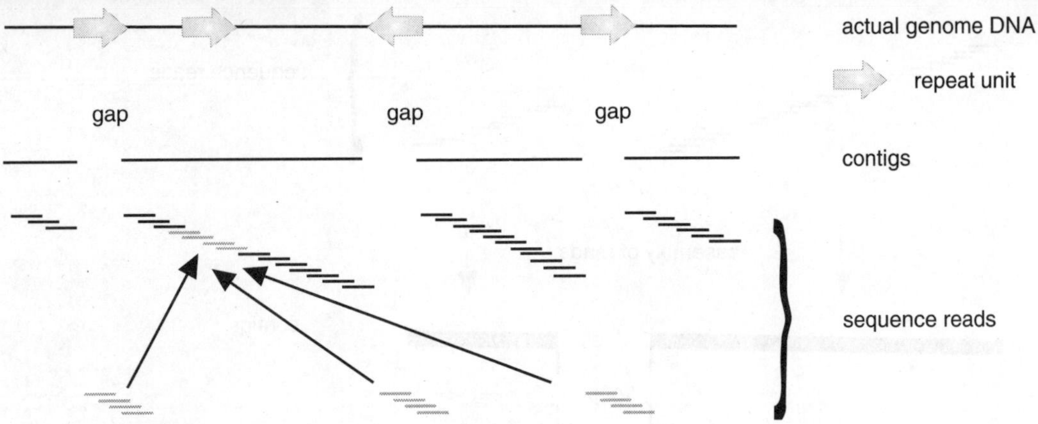

Figure 2.3: Showing the repeats located at different sites on chromosome.

The resulting DNA fragments are *subjected* to DNA sequencing on both strands by primer walking, and gaps are closed one by one. Gaps caused by other *multiple-copy* units can also be closed using these procedures. Some bacterial surface *proteins* may contain internal repeats with long core lengths (encoded by DNA many hundreds of base pairs in size) and each *suspected* open reading frame (ORF) needs careful evaluation for *successful* assembly. ,

Assembling cotigs

After gap closing, it is *necessary* to confirm whether the contigs have been assembled correctly. The final genome *sequence* must *coincide* with actual *restriction* maps obtained by pulsed-field gel electrophoresis (PFGE) using several restriction enzymes. Long-PCR "*walking*" along the whole genome sequence may *compensate* for lack of information on *physical mapping* and *restriction* pattern analysis of long-PCR fragments (appprox 16 kb will further increase the *reliability* of the final sequence.

A high level of *redundancy* is required for optimal DNA sequencing, i.e., each area of the genome should be *represented* several times in the library and hence sequenced several times; up to 10 levels of *redundancy* may be needed to solve the problem in a small *fraction* of abnormal clones.

In the case of the *S. aureus* genome sequence (2.8 Mb), 56,000 sequence reads were *sufficient* to obtain an average 10 levels of redundancy. To increase accuracy, it may be necessary to edit *manually* sequencing errors following visual inspection by using editor srogram.s, such as *Consed*.

PHYSICAL MAPPING & DNA SEQUENCING

An entire genome structure may be determined by physical mapping and DNA sequencing as described above. The next step is to annotate it. Annotation is the determination or prediction of sequences in a genome that correspond to genes, including protein-encoding and RNA-encoding regions. This generally involves the use of computer *programs* that scan genomic DNA sequences to predict coding sequences and assign *functional* roles to *predicted* proteins.

There are many programs freely available for use on personal computers or through *websites*.

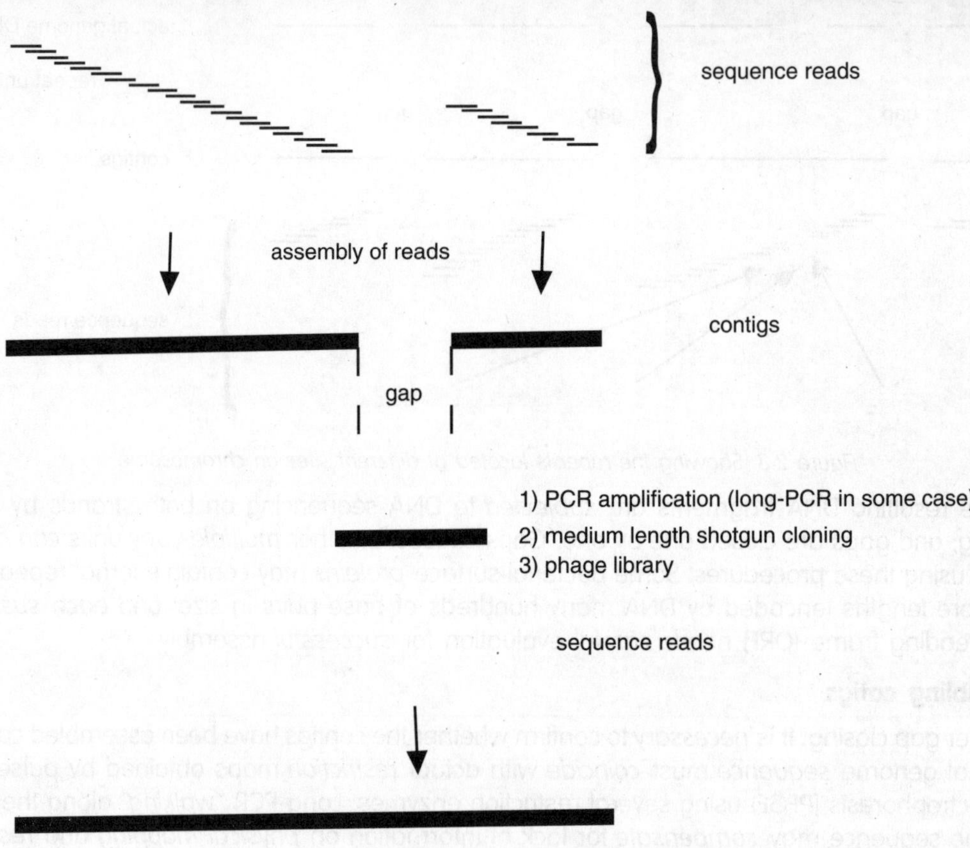

Figure 2.4: Briding Gap by the DNA fragments.

The features of these programs are beyond the scope of this chapter, and we have limited the *discussion* to the *essential* steps required for annotating a newly sequenced *bacterial* genome.

Role of Genetic Code in Determing Nucleotide Sequence

The simplest method of predicting genes encoding proteins is to search the DNA sequence for ORFs. An ORF is a DNA sequence that contains a *contiguous* set of *codons* (i.e., three nucleotides that *specify* a particular *amino acid*). There are six potential reading *frames* in every DNA sequence, with three *possible* frames on-one strand, and another three on the *complementary* strand. All ORFs read in the 5' to 3' direction on a given sequence.

The genetic code used to translate nucleotide sequences into amino acids differs slightly between organisms; four *codons* of vertebrate mitochondrial DNA differ from the universal genetic code. The code that is most *appropriate* for the target species must be *selected* for ORF *prediction*, and the *bacterial genetic* code should be used for *bacterial* genomes.

Bacterial *translation* is initiated by *recruitment* of a 70S *ribosomal complex, initiation factors* (IFs), and *formylmethionine*-tRNA^{Met} (fMet-tRNA^{Met}) fMet-tRNA^{Met} has a higher binding *affinity* for IFs than *methionine*-tRNA^{Met}, and it recognizes not only the universal initiation codon AUG in mRNA, but also GUG and UUG as alternatives.

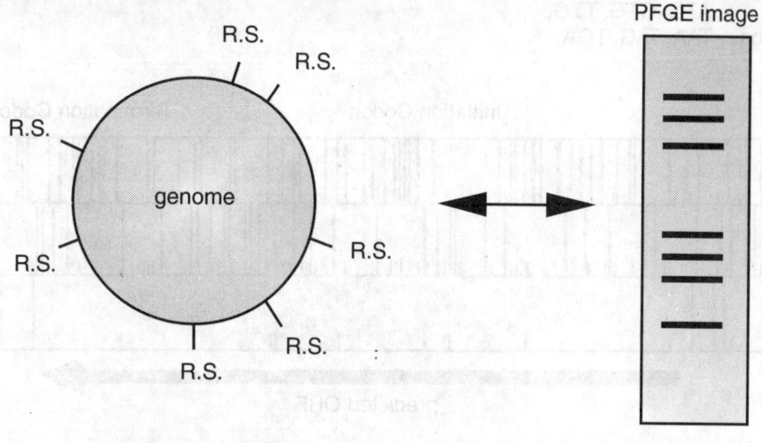

Figure 2.5: Cap restriction enzymes to prove the physical map.

A ribosome-binding sequence (also known as a Shine-Dalgarno [SD] sequence) is sought to evaluate the most probable location of predicted genes. SD sequences comprise *purine-rich* (A and G) sequences with a strong similarity to the complementary sequence near the 3' end of 16S rRNA.

An SD sequence located 6- 1 by upstream of an *initiation* codon is important for *translation initiation* Figure elsewhere in this chapter shows an example of a search of *candidate* SD sequences and initiation colons; the pairs with the most *appropriate* length of spacer region would be the most probable SD sequence and *initiation* site of the ORF.

The order of the codons in a gene determine the order of the *amino acids* in the *protein* encoded by that gene. The order of the codons also reflects the *evolutionary* origin of the gene, and *constraints* associated with its *expression.*

Each species has a characteristic pattern of use of *synonymous* codons; the *preference* of codon usage reflects *species-specific* % GC content and relative absence or *abundance* of some kinds of tRNA species.

See table elsewhere in this chapter the codon usage for *alanine* and *leucine* in the low-% GC Gram-positive bacterium *S. aureus,* the *enteric* gram-negative bacterium *E. coli,* and the high-%o GC Gram-positive bacterium *Mycobacterium tuberculosis. S. aureus* has a low GC content (33.5%) and has a strong codon bias in favour of A and T/U *nucleotides.*

Leucine is encoded by six possible codons, but *S. aureus* prefers UUA and UUG (74% of all six codons), rather than other GC-rich codons (CUA, CUC, CUG, and CUU). On the other hand, *E. coli* and *M. tuberculosis* prefer GC-rich codons; again codon usage and % GC content of the genome correlate with each other.

Computer programs such as GeneMark and GLIMMER use hidden *Markov models* (HMMs) and *interpolated Markov models* (both complex statistical algorithms) to identify potential genes in DNA sequences, based on the *characteristics* of *protein-encoding* sequences described above. Such *programs* have been developed to identify genes without human *intervention.* However, the predictions often include false-positive or *false-negative* ORFs.

Initiation Codon : ATG, GTG, TTG,
Termination Codons : TAA, TAG, TGA,

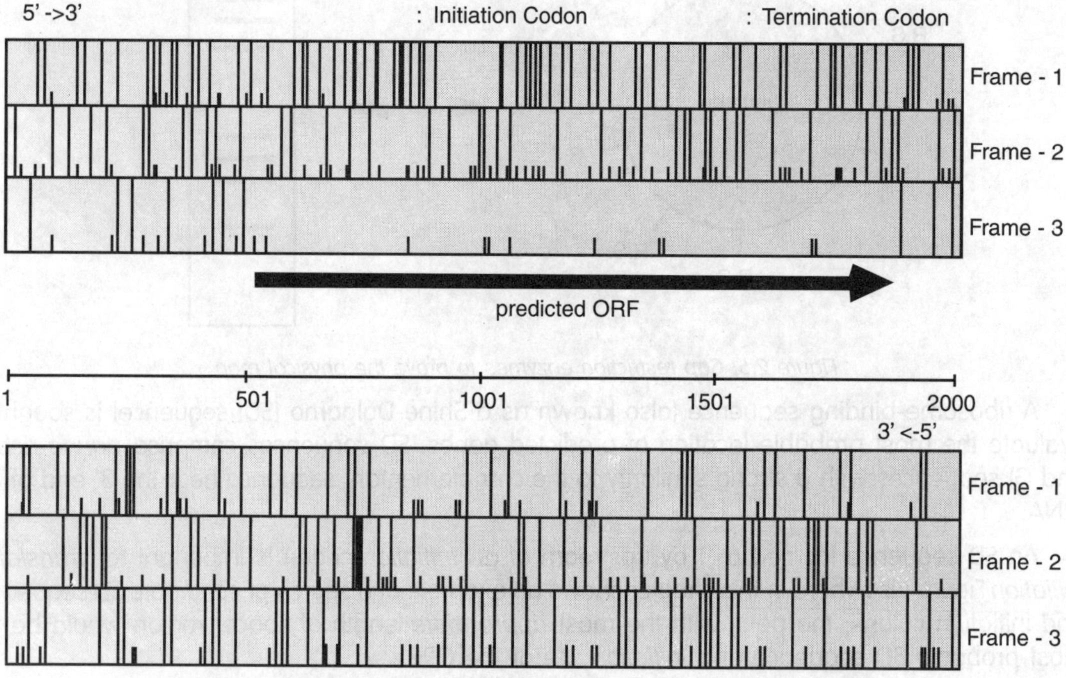

Figure 2.6: Figure showing the initiation & termination codons.

Thus, revision by visual *inspection* may increase the *reliability* of ORF *identification*.

Searching New Genomes

The most reliable way to identify a gene in a new genome is to find a close homolog in another *organism*. Homologs are proteins that share significant *sequence* similarity, which can be *attributed* to a common *evolutionary ancestor*.

Homology *searches* can be done *effectively* using programs such as BLAST to search all the entries in the GenBank database (http://www.ncbi.nlm. nih.gov/BLAST/). Gene *annotation* is based on finding significant *alignment* of *predicted* ORFs to sequences of known function. Gene *products* are classified on the basis of function, *biological role*, an *cellular location*. However, many of the genes in new genomes show no *significant homology* to *GenBank* entries.

Matches of lesser significance *provide* only a *tentative prediction* and should be used as a working *hypothesis* of gene function X. Thus, further analyses of *motifs, structure*, and *localization* can help to assign *characteristics* to *hypothetical proteins*. In *comparisons* between the proteomes of organisms, each protein is used as a query in a database similarity search against another *proteome*, or *combined* set of *proteomes*, in a database such as GenBank, SWISSPROT, or *Protein Information Resource* (PIR).

A pair of proteins in organisms of two species that align along most of their lengths with a highly significant alignment score are likely to be orthologs, i.e., proteins that share a common ancestry and that have kept the same function following an *evolutionary speciation* process. These proteins perform the core *biological* functions shared by all organisms, including DNA replication, transcription, *translation,* and *intermediary metabolism*. They do not include *proteins* unique to the biology of a *particular organism*.

The Clusters of Orthologous Groups (COGS) database may aid assignment of gene function. This database is used by *performing* an all-by-all genome *comparison* across a *spectrum* of *prokaryotic organisms,* and a portion of the yeast *proteome*. Each COG consists of individual orthologous proteins or *orthologous* sets of *paralogs* from at least three *lineages*. All-against-all *self-comparison* may reveal *paralogs* with a high degree of sequence similarity that have almost certainly *originated* from gene duplication events within a species.

Sequence *alignments* found in database *searches* indicate the presence of one or more conserved domains in each cluster or group of clusters. The most similar *regions* in a multiple set of proteins may represent conserved functional or *structural domains*. Hence, such clusters are next analyzed for the presence of known *domains*, by searches of a domain database. Functional searches with global alignments can be performed through the Internet and alignments can be performed with locally installed software, such as

CLUSTALW or CLUSTALX which has a *graphic interface*. The *software* is available from ftp.e i,ac.uk/pub/software. A multiple alignment of a set of sequences may be viewed as an *evolutionary* history of those sequences. If they align well, they are likely to have been derived recently from a common ancestral sequence. The task of aligning a set of sequences of varying *relatedness* is identical to that of *discovering* the *evolutionary* relationships among the sequences.

Multiple Sequence alignment

InterPro is a useful resource for whole-genome *analysis*, and has been used to analyze the *proteomes* of a number of organisms for which the genome *sequences* are known.

```
SA0006 (7005..9674)

6781  TTGGAGAAAT  GAATGCAGAT  CAATTATGGG  AAACAACAAT  GAACCCTGAG  CACCGCGCTC

6841  TTTTACAAGT  AAAACTTGAA  GATGCGATTG  AAGCGGACCA  AACATTTGAA  ATGTTAATGG

6901  TGTGACGTTGT AGAAAACCGT  AGACAATTTA  TAGAAGATAA  TGCAGTTTAT  GCAAACTTAG

6961  ACTTCTAAGC  GCTGTGAACT  GAACTTTTGA  AGGAGGACT   CTTGATGGCT  GAATTACCTC
                                                      M  A  E  L  P  Q
7021  AATCAAGAAT  AAATGAAGGA  AATATTACCA  GTGAAATGCG  TGAATCATTT  TTAGATTATG
       S  R  I  N  E  R  N  I  T  S  E  M  R  E  S  F  L  D  Y  A
7081  CGATGAGTGT  TATCGTTGCT  CGTGCATTGC  CAGATGTTCG  TGACGGTTTA  AAACCAGTAG
       M  S  V  I  V  A  R  A  L  P  D  V  R  D  G  L  K  P  V  H
7141  CGATGAGTGT  ACTATATGGA  TTAAATGAAC  AAGGTATGAC  TACCGGATAAA TCATATAAAA
       R  R  I  L  Y  G  L  N  E  Q  G  M  T  P  D  K  S  Y  N  K
7201  AATCAGCACG  TATCGTTGGT  GACGTAATGG  GTAAATATCA  CCCTCATGGT  GACTCATCTA
       S  A  R  I  V  G  D  V  M  G  K  Y  H  P  H  G  D  S  S  I
```

Figure 2.7: Possible initiation codons (ATG) of gyrA of Staphylococcus aureus N315.

Table 2.1: Proportional distibutiar codons.

Codon	Amino acid	Proportion (%)		
		S. aureus N315 coding GC% (33.5)	*E. coli* K12 coding GC% (51.8)	*M. tuberculosis* H37Rv coding GC% (65.9)
GCA	Alanine	47	21	10
GCC	Alanine	7	27	45
GCG	Alanine	15	36	37
GCU	Alanine	31	16	8
CUA	Leucine	9	4	5
CUC	Leucine	2	10	18
CUG	Leucine	3	50	52
CUU	Leucine	12	10	6
UUA	Leucine	59	13	2
UUG	Leucine	15	13	18

InterPro combines a number of databases (Pfam, PRINTS, PROSITE, ProDom, SMART, TIGRFAMs, SWISSPROT, and TrEMBL) that use different *methodologies*, and a varying degree of *biological* information, on *well-characterized* proteins to derive protein signatures.

This kind of analysis identifies the number and types of domains that are shared among organisms, or that have been duplicated in *proteomes* to produce *paralogs*. Prediction of protein *localization* may assist *functional* assignment of the products of *predicted* ORFs.

The PSORT program predicts the presence of signal sequences that contain an N-terminal basically charged region, and a central hydrophobic region. It also predicts probable *transmembrane* segments and the signal sequences of *lipoproteins*. These predictions can be used to suggest the most *probable* location of the ORF product (*cytoplasm, periplasmic space, extracellular*, or *membrane-associated*).

For example, some surface proteins in *staphylococci* are *anchored* to the *peptidoglycan* chains with a LPXTG amino acid motif that is recognized by the sortase SrtA protein. Finding such a well-characterized motif in an unknown predicted ORF will strongly suggest the *probable* protein *localization*.

Products of predicted genes

The function of predicted ORFs can be determined through *homology* and *motif searches*. The products of predicted genes are classified into five types based upon the extent of their similarity to *sequence* entries in the various *databases searched*.

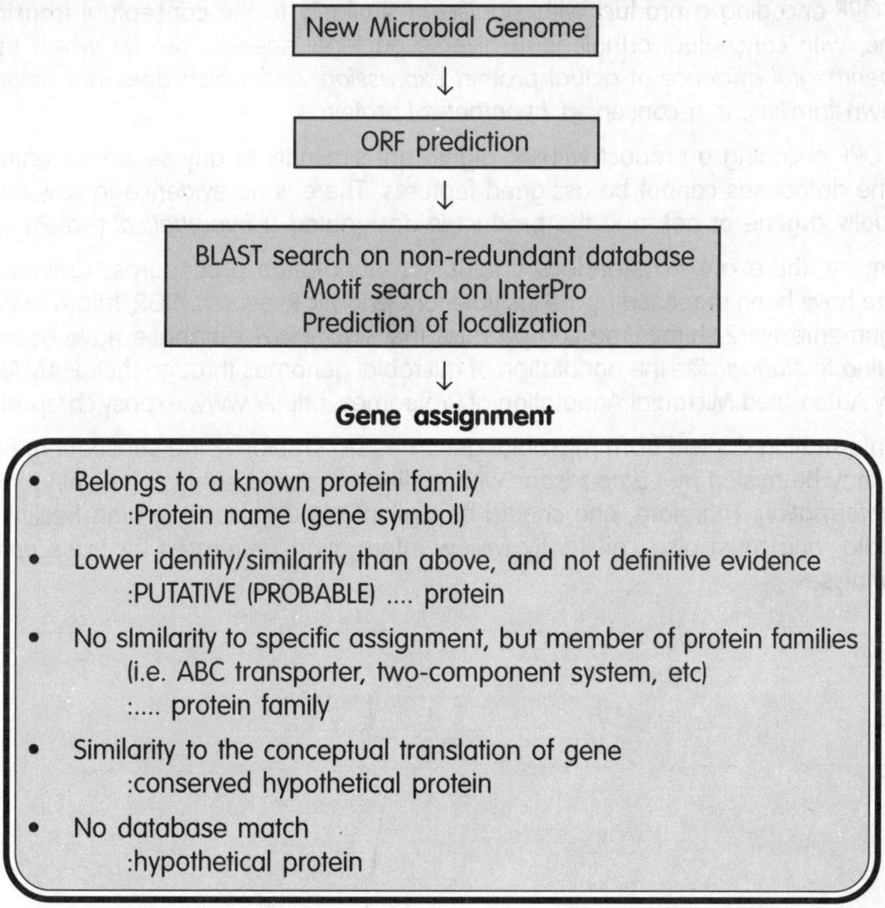

Figure 2.8: Presentation of the stages of genome annotation from ORF prediction to gene assignment.

- An ORF encoding a product with good similarity to a protein of known function, with good sequence alignment along the entire length, and conserved active sites and motif sequences, is usually given the *designation* of the corresponding protein (and gene) of other *bacteria* (commonly *E. coli or Bacillus subtilis).*
- An ORF encoding a product that matches a characterized protein along most of its length but which has a low similarity (less than 40% *amino acid* identity) and no matches with *protein motifs,* is designated a putative protein sharing the putative function of the matched protein.
- An ORF encoding a product for which there are no matches with specific protein entries along the *entire alignment* of the ORF, but which shows strong similarity to a particular protein family (e.g., ABC *transporters* and *transcription regulators* with helix-turn-helix DNA-binding motifs) when analyzed by motif searches, cannot be assigned a specific gene and *protein name.* It is *designated* a *probable* member of the particular protein family.

- An ORF encoding a product with *significant* similarity to the conceptual translation of a gene, with conceptual orthologs in diverse bacterial species, but for which there is no *experimental evidence* of actual protein *expression*, and which does not belong to any known families, is a conserved *hypothetical* protein.
- An ORF encoding a product with no significant similarity to any sequence entries in any of the databases cannot be assigned features. There is no evidence to say whether it is actually a gene or not, and the product is designated a *hypothetical* protein.

In summary, there are no standard criteria for *annotation* procedures. Criteria for gene nomenclature have been proposed by the Institute for Genomic Research (TIGR, http://www.tigr.org/ER/db assignmentextver2.shtml). The compilers of the SWISSPROT database have been verifying and attempting to *standardize* the annotation of microbial genomes through their HAMAP initiative (High-quality Automated Microbial Annotation of Proteomes, http:// www.expasy.ch/sprot/hamap).

As the information derived from microbial genomes *accumulates*, the subsequent annotation *procedures* may be misled by *comparisons* with earlier annotated entries containing incorrect or surpassed *information*. Therefore, one should be vigilant about producing high-fidelity genome-sequence data, and must also constantly review information *generated* by *transcriptome* and *proteome* analyses.

Chapter 3

Sequencing Genome

All organisms and even *prokaryotic* and *eukaryotic* cells have definite genome. In the majority of cases the sequencing of genomes provides biologists with *invaluable datasets* for investigating the evolution of individual species, differences and *similarities* between various species, and functional *characteristics* of genomes. *Comparative* analysis of *genomes* makes an important contribution to solving these and many other problems.

In most cases, this analysis is based on the *alignment* of genomic sequences followed by *investigation* of the level of conservation and the search for sequence signals specific to a particular genomic function. There are several approaches to each step of such studies, but regardless of the particular approach, there is a need to visualize the results of this *comparative* analysis.

Alignment is probably the most investigated area of *computational* biology, but it is still a subject of intensive work by many groups. There are several types of *pair-wise* alignments, i.e., global, local, or a *combination* of *global* and *local*, described in detail elsewhere.

The availability of several *assemblies* of large genomes made possible the development of *whole-genome alignment techniques*, which *generated* a number of *precomputed* alignments that are available to the *community*.

All techniques are unified by the common principles of finding the most similar genomic *intervals* (*anchors*) followed by extending these regions and *chaining* alignments to make them *contiguous*. The basepair level of visualization of alignments provides *investigators* with the most detailed comparative data, the same holds true for *multiple* alignments.

At the larger scale, visual presentation of *rearrangements, inversions, gap composition*, and order of *fragments* of a draft sequence in the alignment are important for understanding the *biology* of a particular *genomic* interval.

One of the main purposes of *comparative* genomics is to provide a detailed analysis of *conservation* among orthologous intervals in different species. Defining which genomic intervals have been subject to negative (*purifying*) selection can bring us closer to understanding functions of different genomic elements. Methods for calculating *conservation* in alignments range from a simple window-based approach in PipMaker and VISTA to the *phylogenetic* hidden Markov model *Phastcons,* another *statistical model,* Gumby *visualization* of sequence conservation is a critical aspect of comparative sequence analysis because manual *examination* of alignment on the scale of long genomic regions is highly *inefficient.*

This is why *alignment-browsing* systems are specifically designed to identify *well-conserved* segments. Different methods for calculating segments of conservation define the type of *visual* presentation, for example PIPMakerV represents the level of *conservation* in *ungapped* regions of BLASTZ local alignment as *horizontal dashes;* VISTA and SynPlot display *comparative* data in the form of a curve, where conservation is calculated in a sliding window of a *gapped* global alignment; PhastCons also generates a *contiguous* curve and *Gumby* scores are presented as the histogram-like Rank VISTA plot.

Internet-based genome *browsers,* emerging relatively recently, present the most essential tools for *investigating* genomic sequences because they integrate all sequence-based *biological* information on genes or *genomic regions.*

They are easy to use and very *efficient* n retrieving large amount of relevant biological data. UCSC Browser,Ensembl and MapView at National Center for Biotechnology Information provide *comprehensive* data related to a number of *vertebrate, invertebrate,* and other *genomes.* In contrast, VISTA Browser is highly *specialized* and was built to show the results of comparative *analysis* of genomic sequences based on DNA alignments, both *whole-genome* and *interval-based.*

Here, we present this *computational* tool with all the *internal* and *external extensions* and *demonstrate* its capabilities by analyzing several genomic intervals. VISTA presentation of comparative data is easy to interpret both on a small and a large scale, i.e., at different levels of *resolution.*

All VISTA programs and servers use the same type of *visualization,* making interpretation of alignments easy. Because VISTA tools are being *constantly* improved and enhanced, new options and *capabilities* can be found on the website. The VISTA support group will help users explore these new options and answer questions.

WHOLE-GENOME ALIGNMENTS

Whole-genome alignments accessible through VISTA Browser are based on the local/global approach developed in the group. These *alignments* are available for a number of *vertebrates, invertebrates, plants,* and *others species.*

The list of whole-genomes alignments is constantly being updated by the VISTA group when new assemblies become available. Results of VISTA *comparative* analysis are also available for a number of bacteria. *Precomputed* full scaffold alignments for *microbial genomes* are presented as a component of Integrated Microbial Genomes developed in the Department of Energy's Joint Genome Institute, and are also *available* through the VISTA portal.

Accessing Browser

As any other genome browser, VISTA Browser provides a view of a particular interval of a base genome. Thus, as the first step, the user needs to choose a *genomic* interval on the selected base genome. Access the VISTA portal page online at http://genome.lbl.gov//vista and click the "VISTA Browser" link in the "*Precomputed* whole genome alignments" section, or use the direct link to the VISTA Browser gateway (http://pipeline.lbl.gov). Detailed help pages are available online (http:// pipeline. lbl.gov/help.shtml). Select the "Base genome" from the pull-down menu on the left. *Base genomes* are identified by the name of a species and a date of assembly.

After the Base genome is selected, a list of all available genome for this *alignments* will appear on the gateway page. Define a position on the base genome. The user can input a position on a chromosome or a contig, as well as supply a gene name. The gene name should correspond to the *annotation* datasets used for a particular base genome.

The gateway page describes which annotation are used for each base genome in the browser, i.e., RefSeq for human, mouse, and *Drosophila melanogaster*, FlyBase for *D. melanogaster*, *TIGR* annotation for rice, and others. An example of an input is shown in Figure elsewhere in this chapter, where *D. melanogaster* is selected as the Base genome, and an arbitrary interval, chr2L:816,000-828,000, is selected as the Position.

The user can choose either "VISTA Browser" or "VISTA *tracks* on UCSC Browser" as methods to view the results. Description of the differences between them will follow. VISTA Browser requires Java software to be installed on the computer. If the user entered a *chromosome/contig* position or the name of a gene with a unique match, selecting "*Go*" will take the user directly to the browser. If a gene name is entered without a unique match, the user will be directed to a page that lists all *entries* that contain the search term.

How to Display Browser

The display consists of three main sections: a Control Panel on the left hand side, the central browser window(s), and a *horizontal toolbar* at the top. Here, we describe what these three sections consist of and how to use them.

Control Panel

The main functions of the Control Panel are illustrated in the Figure. The window that appears on the *desktop* of the computer when the browser is accessed through the gateway at http:// pipeline.lbl.gov. This part is also well illustrated in the figure.

The conservation plot displayed on the right is based on the alignment of the base genome *D. melanogaster* with the genome of *Drosophila pseudoobscura* the second species that is indicated below the plot on the right).

In the section with the five pull-down menus on the left, the name of the base genome can be seen, *position* on the *genome*, the *annotation* track used in the display, and the number of rows in the plot display ("*Auto*" is a default). Each of these menus provides the user with a choice of options, for example, a user can replace the RefSeq annotation track with the FlyBase annotation track.

Selecting "1" as the number of rows changes a three-row *continuous* view of the genomic interval to a one-row view. Next, the "select/add" menu allows the user to view what other alignments are available for the *D. melanogaster* genome. Selecting *Drosophila simulans* in this menu will open a small window that allows the user to choose display parameters for the plot of the alignment of *D. melanogaster* and *D. simulans*. After changing the parameters or using the default parameters, clicking OK will cause the browser to display conservation for t o alignments on the same interval of the base genome . Figure elsewhere in this chapter shows the browser display after adding two more VISTA windows, the *D. yakuba* and *D. ananassae* alignments to the base genome.

Among the choices in the select/add menu, will be the RankVISTA plots for some of the alignments. Rank VISTA is an alternative way of scoring conservation in alignments that could be useful in some *applications*.

In the Information section on the left are the coordinates of the cursor on the base genome and the name of the *chromosome* or contig of the second species aligned in this position. This name displayed is for a selected plot, or for the default *alignment* if no plot is selected. If the displayed genomic interval has masked repeats, the Color Legend box indicates how different kinds of repeats are displayed above the plot.

Pair Wise Alignment

The VISTA conservation window or several stacked windows occupy a central position in the Browser. Conservation is displayed in a standard VISTA format of *peaks* and *valleys*, and the height of each peak is indicative of the level of *conservation* in this area. The *horizontal* bar on the top of the central section depicts the length of the entire *chromosome* and shows the location of the *investigated* interval on this chromosome.

Arrows on the top of the plots show the position and direction of genes, with their exonic intervals in blue and UTRs in turquoise, according to a selected *annotation*. Thus in VISTA plots, peaks depicting conserved *sequences* (CNSs) are blue if they are in exonic intervals of the base genome, *turquoise* if they overlap with UTR, or red for all unannotated *sequences*, i.e., *intronic, intergenic*, or without clear *assignment*.

The bar below the plot is gray for *continuous uninterrupted* alignment, red where several intervals of the second genome are aligned to the same interval of the base *genome* (overlap, at chr2L:823,000-825,000 interval of *D. melanogaster/D. simulans* alignment) or where the alignment is interrupted (for example chr2L:824,200-826,500 interval in the same alignment).

Holding the left mouse button down and selecting an area on the base genome allows for zooming in on the interval of interest. Left-clicking any plot selects it, and that selection is necessary for a number of *manipulations* described next. Selected plots are *shaded* gray.

TOOLBAR CONTROL

Different control options are available either through the Toolbar, or a menu at the top of the Browser. Keeping the cursor over any of the buttons in the *Toolbar* shows a description of the option.

The buttons are:

Add VISTA Curve: works the same way as "select/add" menu in the Control Panel.

Remove VISTA Curve: one of the curves should be selected to use this option.

Save as: displays a window with a selection of formats (pdf, jpeg, or gif) for saving the plots to a file.

Print.

Scroll backwards and forward on the base genome. Zoom in and out.

Return to previous and next position on the base genome.

Browsers: link to the same interval on the base genome displayed in the alternative browser(s). For some genomes, this button will bring up the UCSC browser with additional VISTA curves/ control options. *Relevant browsers* also include the JGI browser for a number of species, RGD for the rat *genome,* and others.

To use the following three buttons it is necessary to select one of the plots: Alignment details (1): gives access to a page with detailed comparative information, also referred to as *"Text Browser."*

Alignment: shortcut to a text file with an alignment.

Curve parameters: opens a window for changing *conservation* parameters used for building the VISTA plot, the same as the window in Figure elsewhere in this chapter.

Right-clicking on the curve opens a selection window that gives access to some of the options of the *Toolbar* (*Details, Parameters, Alignment, Add/Remove*), with an additional option of changing the base genome.

TEXT BROWSER

This page links the alignments to other sequence-based information. The user will find the coordinates of conserved regions, their sequences, annotations, and other available data. Figure elsewhere in ths chapter shows the most basic set of options in the "Text Browser," obtained from the VISTA plot of *D. melanogaster vs D. ananassae.*

The names of *participating genomes* as well as the program used for the alignment are shown in the top banner. Below the banner are the coordinates of the currently displayed region and a link back to VISTA Browser, an alternative browser (VISTA Tracks on UCSC in this case), and a pull-down menu with a choice of *annotation.* Links in the next row give access to the *coordinates* of annotated genes in the *interval,* as well as the *coordinates* of CNSs.

The user will notice that when the conserved *regions* are displayed, their lengths are actually web links. Clicking on the links will bring up the conserved sequences from both of the *participating* organisms. In the main table listed next, each alignment generated for the base *organism* is displayed. Columns, except for the last one, refer to the sequences that participate in the alignment.

The last column contains detailed information on the whole alignment. Each row is a separate alignment, and displays pairs of genomic intervals of the two organisms participating in this alignment. The first cell of each row contains a small image of the VISTA plot of this alignment, which is helpful when *several* alignments are compared for an interval and the user wants to evaluate relative quality of those including *alignment* overlaps. *"Sequence"* links to a FASTA-formatted DNA segment that participates in the *alignment.*

Clicking on the "VISTA Browser" link will launch the browser with the associated species as the

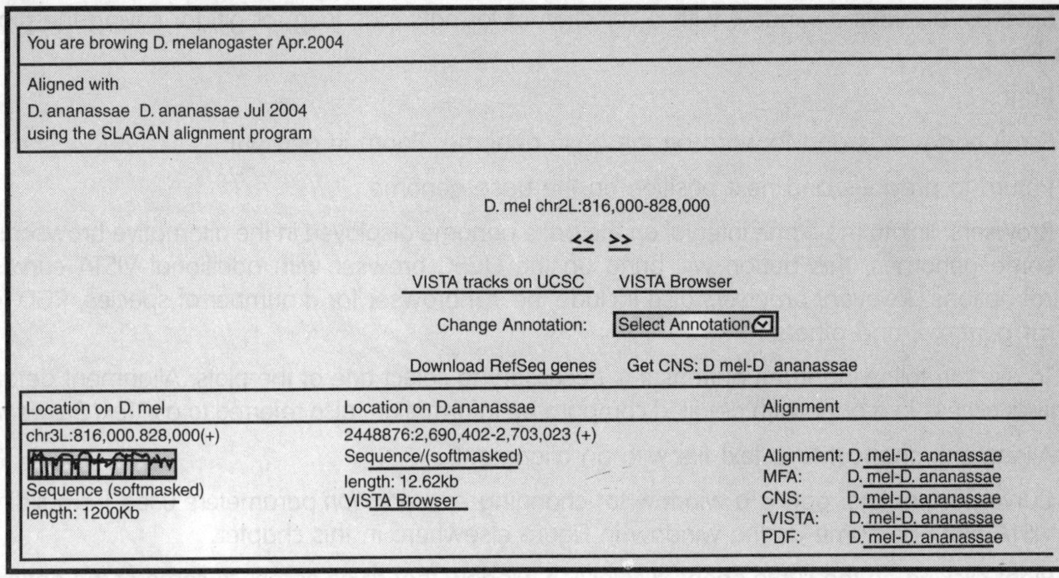

Figure 3.1: Detailed information display provides access to the data underlying the VISTA graph of the genomic interval chr2L:816-828000 of D. melanogaster aligned with D. ananassae.

base. The last column provides links to the alignments in *different* formats, a list of conserved regions from this *alignment*, and links to static pdf-formatted plots of this alignment.

Special Cases of Alignment

Text Browser design allows for flexibility in presenting information relevant to participating sequences and their alignment. Next are several special cases:

1. When the Shuffle-Lagan program is used for comparing user-submitted *sequences* or *microbial* genomes, there will be a link to dot-plots of the alignments produced.
2. When several intervals of a second species are aligned to a particular interval of the base genome with or without overlap, the first column will display several VISTA pictures for each subinterval of the alignment.
3. In case of a multiple alignment, there will be more than one column with the data on the aligned to the base genome species. Each column will provide details on a *particular* organism.
4. If the *examined* region of the base genome is shorter than 20 kb, Text Browser will provide a rVISTA link to start this analysis.
5. If the examined region is long enough for the Rank VISTA evaluation of *conservation*, the link to this tool will be found in Text Browser.

If Text Browser displays new links not described in this chapter, Help pages will provide detailed *description* of these modules.

VISUALIZING ALIGNMENTS

VISTA Browser has been built to visualize alignments of any length, thus in addition to *displaying comparison*.of the whole genomes it is used for comparative *analysis* of user-submitted sequences. VISTA portal (http://genome.lbl.gov/vista) offers a choice of several automatic servers described briefly next. More details on the VISTA servers are available in our *previous publications*.

VISTA pages also provide extensive help on selecting a type of analysis and finding optimal *parameters* for a *particular* project. In Genome VISTA, a single sequence (*draft* or *finished*) is compared with whole genome *assemblies*.

For a submitted sequence, the server finds candidate *orthologous* regions on the base genome, and provides detailed comparative analysis. mVISTA is designed to perform pair-wise or multiple alignments of DNA sequences from two or more species up to megabases long and to visualize these alignments together with their annotations.

Depending on the project, a user can choose one of the three alignment programs: AVID for global pairwise and multiple pair-wise alignment (one of the sequences can be in a draft format), LAGAN for global pair-wise and *multiple* alignment of finished sequences, or Shuffle-LAGAN for global alignment with synchronized detection of rearrangements and inversions. rVISTA (regulatory Vista) combines searching the major transcription factor binding site database TRANSFAC™ *Professional* from Biobase with a comparative sequence analysis. It can be used directly or through links in mVISTA, Genome VISTA, or VISTA Browser. Phylo-VISTA allows a user to visualize *submitted* multiple sequence alignment data wile taking the *phylogenetic relationships* between sequences into account.

POINTS TO REMEMBER

1. How to install Java. VISTA Help section provides a detailed instruction on this installation (http://pipeline.lbl.gov/vgb2/help/Java win instructions.shtml). The latest version of J2SE from the Java download page of Sun Developer Network will be needed (http://java.sun.com/j2se/1.4.2/download.html).

2. How VISTA curves are calculated. The Vista curve is calculated as a windowed-average identity score for the alignment. A variable sized window (Calc Window) is slid across the alignment and a score is *calculated* at each base in the coordinate sequence. That is, if the Calc Window is 100 bp, then the score for every point X is the percentage of *exact matches* between the two *alignments* in a 100-bp wide window centered on that point X.

   Because of resolution constraints when visualizing large alignments, it is often necessary to condense *information* about 100 or more *basepairs* into one display pixel. This is done by only *graphing* the *maximal* score of all the basepairs covered by that pixel.

3. How to choose display *parameters*. The *parameters* selected for visualization of alignments have a significant effect on the VISTA results. A user can vary the following *parameters*. (1) a window for calculating the VISTA curve; (2) window size for finding CNSs (Min Cons Width); (3) percent of identical *nucleotides* in the window for finding CNSs (Cons Identity); (4) minimum level of Cons Identity shown on the plot (Minimum Y); (5) maximum level of Cons Identity shown on the plot (Maximum Y).

   Parameter (1) defines smoothness of the plot, selection of parameters (2) and (3) depends on the similarity of compared sequences. The default parameters of 100 by for a window and 70% for *similarity* normally need to be reduced for distant species with lower level of conservation, and increased for higher than human/mouse similarity. Generally it takes several trials to retrieve CNSs with meaningful level of *conservation*. In many cases, precomputed Rank-VISTA provides an additional list of highly conserved elements calculated by a different technique. Rank-VISTA parameters are also *adjustable*, and their *description* can be found in the Help section.

Chapter 4

Nucleotide Polymorphisms of Genomic DNA

The identification of nucleotide polymorphisms between closely related organisms is valuable as these polymorphisms can serve as a basis for evolutionary and population analyses. Such nucleotide substitutions may also result in the replacement of amino acids in encoded proteins and, therefore, may underlie important phenotypic differences, such as virulence or tissue tropism.

However, because sequencing numerous genomes is currently neither a rapid nor fiscally practical approach, DNA microarrays provide a useful alternative as they can survey and identify changes across an entire genome. DNA microarrays are competition-based assays in which a reference DNA sample and a test DNA sample are fluorescently labeled before hybridizing with immobilized DNA.

Because the reference DNA is identical to the immobilized DNA, the ratio of the fluorescence at each locus is dependent on the nucleotide sequence of the test isolate. An equivalent signal ratio indicates uniform hybridization and signifies a very high nucleotide identity; the more variable the sequence of the test DNA at that locus, the larger the ratio becomes.

This is a qualitative difference that loses its correlation when two sequences at a region fall below ~75% identity. The microarray assay described here involves (1) the detection of loci on the array that may be biased, (2) demonstrating that there is a correlation between the signal ratio and percent nucleotide identity of two sequenced genomes, and (3) testing samples to identify those regions that are variable.

This procedure was conducted on 14 human serological variants of *Chlarnydia trachomatis,* which are nearly identical on the genomic level but are specific for one of three distinct tissue tropisms and pathologies. Analyses revealed genes with nucleotide polymorphisms, some of which may be associated with the organism's fitness and success within a particular host niche.

REQUIREMENTS

Cultural Conditions

1. HeLa229 cells.

2. T150 tissue culture flask.
3. RPMI Media (Invitrogen, Carlsbad, CA).
4. Fetal bovine serum (Hyclone, Logan, UT).
5. Vancomycin (Sigma, St. Louis, MO).
6. Renografin (E.R. Squibb and Sons, Princeton, NJ).
7. Sucrose phosphate glutamate buffer, pH 7.4: 220m*M* sucrose, 3.83m*M* KH_2PO_4, 8.60m*M* Na_2HPO_4, 4.26m*M* glutamic acid.

Obtaining Genomic DNA

1. TE buffer, pH 7.6: 0.5 *M* Tris-HCl, 0.5 *M* ethylenediamine tetraacetic acid.
2. 10% (w/v) Sodium dodecyl sulfate (SDS).
3. 5 *M* Sodium chloride (NaCl).
4. 20 mg/mL Proteinase K.
5. Cetyltrimethylammonium bromide (CTAB)/NaCl solution: 10% CTAB (Sigma) in 0.7 *M* NaCl.
6. 24:1 (v/v) Chloroform/isoamyl alcohol.
7. 25:24:1 (v/v/v) Phenol/chloroform/isoamyl alcohol.
8. 70% Ethanol (EtOH).
9. Microcentrifuge (Eppendorf, cat. no. 5415D, or equivalent).
10. Isopropanol.

Method of Slide Preparation

1. Premium plain glass slides (Fisher Scientific, Pittsburgh, PA).
2. Slide rack and staining dishes (Fisher Scientific, Shandon Lipshaw, cat. no. 121).
3. Poly-L-lysine (Sigma).
4. Sodium hydroxide (NaOH).
5. 95% EtOH.
6. Desiccator cabinet (Fisher Scientific).
7. Phosphate buffered saline.

Technique of Fabrication

1. 96-Well PCR plates (MJ Research, South San Francisco, CA).
2. 96-Well thermocycler (Tetrad, MJ Research).
3. PCR reagents: Taq polymerase, 10X PCR buffer, 10 m*M* dNTPs, custom oligonu-cleotide primers.
4. Agarose (molecular biology grade).
5. EtOH/sodium acetate solution (3 *M* Na-acetate in 95% EtOH).

6. Table top centrifuge with plate adapters (Eppendorf, cat. no. 5810 or equivalent).
7. 70% EtOH.
8. 96-Well Speed Vac (Savant SC250, Thermo Electron Corporation, Milford, MA).
9. Aluminum sealing tape for Microtiter plates (Fisher Scientific).
10. 384-Well Genetix print plates (USA Scientific, Ocala, FL).
11. 20X Sodium chloride-sodium citrate (SSC) buffer, pH 7.0: 3 *M* Na_3 citrate, *3 M* NaCl.
12. 3X SSC buffer, pH 7.0 (prepared from 20X stock).
13. Microarrayer (Biorobotics Microgrid or equivalent, Genomic Solutions, Ann Arbor, MI).

Postprocessing Slides

1. Slide rack and staining dishes (Fisher Scientific, Shandon Lipshaw, cat. no. 121).
2. Platform shaker.
3. Humid chamber (Sigma, cat. no. H6644).
4. Table top centrifuge with plate adapters (Eppendorf, cat. no. 5810 or equivalent).
5. 1 *M* Sodium borate (Na-borate), pH 8.0, filtered.
6. 1 -Methyl-2-pyrrolidinone (Sigma).
7. Succinic anhydride (Sigma).
8. 95% EtOH.
9. Ultraviolet (UV) crosslinker (Fisher Scientific).

Labeling

1. RadPrime DNA Labeling System (Invitrogen).
2. 10X dNTP Mix (1.2 mM dATP, dCTP, dGTP, and 0.6 mM dTTP).
3. Cy3-dUTP (25 nmol) and Cy5-dUTP (25 nmol) (Amersham, Piscataway, NJ).
4. TE buffer, pH 7.4 (filtered).
5. Microcon 30 (Amicon) filters (Fisher Scientific).

Hybridizations

1. 10 mg/mL tRNA mix (Sigma).
2. Cover slips (Fisher Scientific).
3. 20X SSC, pH 7.0.
4. 2% (w/v) SDS.
5. Wash Solution 1: 1X SSC + 0.05% (w/v) SDS (1X SSC prepared from 20X stock).
6. Wash Solution 2: 0.06X SSC (prepared from 20X stock).
7. Elmer's rubber cement.
8. TE buffer, pH 7.4.

9. Glass staining dish and rack (Weaton, Fisher Scientific).
10. Gene Machines HybChamber (Genomic Solutions).
11. GENEPIX Scanner 4000A (Axon Instruments, Foster City, CA).
12. GenePix Pro 4.0 software (Axon Instruments).

PROCEDURE

Culture of Bacteria

The genome of C. *trachomatis* strain D/UW-3 is fully sequenced and was used to create the microarray. Human-specific strains representing the major serological variants of *C. trachomatis* were selected for analysis to identify differences among those strains that infect one of three specific tissue niches (ocular, urogenital, or lymphatic).

Also included in the study was the murine-specific mouse pneumonitis (MoPn) strain as it is closely related to the human-specific strains and its genome has been fully sequenced.

1. Propagate C. *trachomatis* strains (A/Harl, B/TW-5, BalApache-2, C/TW-3, D/UW-3, E/Bour, F/IC-Cal 3, G/UW-57, H/UW-4, I/UW-12, J/UW-36, K/UW-31, L1/440, L2/434, L3/404, MoPn) in HeLa229 cell monolayers in T-150 flasks containing RPMI medium supplemented with 10% (w/v) fetal bovine serum and 50 µg/mL vancomycin.
2. Isolate chlamydial elementary bodies (EB) by sonic treatments of cell suspen-sions and purify by ultracentrifugation over 30 and 30/44% (w/v) discontinuous Renografin gradients.
3. Freeze 100 µL aliquots at -80°C in sucrose phosphate glutamate buffer until ready to use.

Obtaining Genomic DNA

1. Thaw chlamydial EB samples on ice.
2. Centrifuge at > 13, 000g for 5 min to pellet chlamydial EB samples and discard the supernatant.
3. Wash pellet three times with TE, pH 7.6.
4. Resuspend pellet in 1020 µL TE, pH 7.6 and vortex to mix samples thoroughly.
5. Split sample into two Eppendorf tubes (each containing 510 µL resuspended EB in TE, pH 7.6) and add to each tube: 60 µL 10% (w/v) SDS, 30 µL 20 mg/mL proteinase K.
6. Mix thoroughly and incubate for 4 h at 50 °C.
7. Add 100 µL 5 M NaCl, mix thoroughly, and incubate at 50 °C for 2 min.
8. Add 80 µL CTAB/NaCl solution, and incubate at 65 °C for 2 min.
9. Add 780 µL phenol/chloroform/isoamyl alcohol solution, mix thoroughly, and centrifuge at > 13, OOOg for 5 min.
10. Remove aqueous viscous supernatant to a fresh tube, add an equal volume of chloroform/ isoamyl alcohol, mix thoroughly, and centrifuge at >13,000g for 5 min.
11. Transfer supernatant to new Eppendorf tube and add 0.6 vol isopropanol to precipitate DNA.

12. Centrifuge at > 13, OOOg for 10 min.
13. Discard supernatant, Air-dry DNA pellet and resuspend in 100 μL nuclease-free dH_2O.

Slide Preparation

1. Prepare NaOH Solution:
   a. Dissolve NaOH in dH_2O: 175 g/700 mL.
   b. Stir until completely dissolved.
   c. Add 1050 mL 95% EtOH.
   d. Stir until completely mixed.
   e. If solution remains cloudy, add water until clear.
2. Place slides in metal slide racks (30 slides/rack). Do not use defective slides.
3. Soak slides in the NaOH/EtOH/dH_2O solution for 2 h with gentle rotation.
4. Rinse slides extensively with dH_2O:
   a. Rinse each unit (slide/rack/container) vigorously with dH_2O for 5 min.
   b. Place slide racks in a large clean glass container, and tilt the container slightly for constant water flow.
   c. Wash under running water for 30 min.
   d. Do not let slides dry at any time.
   e. It is critical to remove all traces of NaOH/EtOH.
5. Prepare poly-L-lysine solution in plastic container.
   a. Mix together 100 mL phosphate buffered saline, 800 mL ddH_2O, l00 mL poly-L-lysine.
   b. Bring up the volume to about 1050 mL with ddH_2O to submerge up to three racks of slides.
   c. Mix and split the solution into three plastic containers.
6. Soak the slides in poly-L-lysine solution for 1 h with gentle rotation. Poly-L-lysine adheres to glass, therefore be sure to use a plastic container.
7. After the poly-L-lysine coating, rinse slides by gentle plunging up and down in two different changes of water.
8. Using a table top centrifuge, centrifuge at 72g for 5 min. Be sure to place paper towels below rack to absorb liquid.
9. Store slides in a desiccator for 3 wk prior to use. Slides that are older than 3 mo may result in faint printing and higher background.

Technique of Fabrication

1. Choose primer pairs to amplify all open reading frames (the genome sequence of *C. trachomatis* strain D/UW-3 has 875 open reading frames. Ensure that each primer is between 20 and 23 bases long and produces a PCR product between 150 and 800 bases.

Primers should be produced in 96-well format.

2. Make 5 μM working stock plates of the forward and reverse primers.
3. Set up PCR reactions in a 96-well format using the following concentrations per 20μL reaction: 2 U Taq Polymerase, 10 ng genomic DNA, 1 X PCR buffer, 0.5 mM dNTPs, 0.4 μM custom primer pair.
4. Amplify all open reading frames using the following PCR cycle conditions. Some annealing temperatures may need to be modified to optimize specific PCR products: 94'C for 1 min; 30 cycles of 94 °C for 40 s, 52 °C for 40 s, 72 °C for 1.25 min; 72 °C for 5 min; 4 °C hold.
5. Run all 96-well reactions on a 1 % (w/v) agarose gel and document results.
6. Add 150 μL EtOH/Na-acetate solution into each well.
7. Cover plates with aluminum sealing tape and store in -20 °C freezer overnight to precipitate.
8. Using a table top centrifuge, centrifuge at 3200g for at least 2 h at 4 °C.
9. Dump solution and pat dry on a napkin.
10. Fill each well with 100 μL 70% EtOH and centrifuge for another hour at 3200g.
11. Dump EtOH solution and dry plates using a 96-well SpeedVac or let plates air-dry.
12. Resuspend PCR products in 40 μL $_{H20.}$ Keep samples at 4 °C for at least 18 h, then transfer products to 384-well printing plates.
13. Dry plates down using a SpeedVac or let air-dry and resuspend in 12 μL 3X SSC.
14. Let plates resuspend at least overnight before printing.
15. Using a Microarrayer, spot DNA from all 384-well plates onto prepared poly-L-lysine treated slides. To provide duplicate assessments for each hybridization experiment, each PCR product should be printed twice per slide when possible.

Postprocessing Slides

1. Before starting:
   a. Warm a heat block to 90 °C and invert the insert.
   b. Place slide-staining chambers in front of the heating block.
   c. Fill slide staining chamber with H_2O.
   d. Get a dust free board needed for cross-linking step.
   e. Place the following items in a chemical hood: large glass beaker with ~1 L boiling H_2O (cover with foil while heating), stirring plate for preparation of blocking solution, and a shaker.
   f. For the preparation of blocking solution, have a clean 500 mL beaker ready.
   g. Have the following chemicals ready: 6 g succinic anhydride, 350 mL 1-Methyl-2-pyrrolidinone, 15 mL 1 *M*Na-borate, pH 8.0 (filtered).
2. Rehydrate the slides:
   a. Rehydrate slides by inverting (array side down) them over H_2O in a slide-staining

chamber. Let the spots become glistening in appearance.

b. Immediately flip slide (array side up) onto a heating block (with inverted insert) and watch the steam evaporate. When the array spots dry in a rapid wave-like pattern, remove slide from heating block. This usually takes about 5 s. Do only one slide at a time.

3. UV cross link slides:
   a. Place slides, array side up, on a flat and dust free board that fits into a UV crosslinker.
   b. Irradiate with 600 μJ UV light.
4. Block free lysines (wear lab coat when working with 1-Methyl-2-pyrrolidinone):
   a. Add 6.0 g succinic anhydride into 350 mL 1-Methyl-2-pyrrolidinone.
   b. As soon as the solids dissolve, quickly add 15 mL of I MNa-borate, pH 8.0, and pour the mixed solution into a slide washing tray.
   c. Quickly place slides into the succinic anhydride solution and vigorously plunge repeatedly for 60 s (do not pour solution over slides).
   d. Turn platform shaker on and rotate vigorously for 15 min.
   e. Remove the slide rack from the organic reaction mixture and place it immediately into the boiling water bath for 90 s.
   f. Transfer the slide rack to the *95%* EtOH wash.
5. Immediately spin dry slides by centrifugation at 72g for 2 min.
6. Transfer slides to a dry slide box and store in a desiccator.

Labeling

Genomic DNA from each strain tested is randomly primed with Cy-5- or Cy-3-labeled nucleotides using the RadPrime DNA Labeling System. D/UW-3 genomic DNA was used as the reference DNA for all hybridizations and labeled with Cy5, whereas all other sample DNA was labeled with Cy3.

1. To an Eppendorf tube, add 0.2 μg genomic DNA and appropriate amount of water to bring volume up to 20μL.
2. Add 20 μL of 2.5X RadPrime Buffer.
3. Boil sample for 5 min and snap cool on ice for 5 min.
4. Keep tubes on ice and add the following to each reaction in order: 5 μL 10X dNTP Mix, 3 μL of appropriate Cy3- or Cy5-dUTP, 2 μL Klenow Fragment (40 U/μL).
5. Incubate reactions for 10 min at 25 °C and then 90 min at 37 °C.
6. Add 5 μL Stop Buffer (0.5 M ethylenediamine tetraacetic acid).
7. Add the two labeled-DNA samples to be compared (appropriate Cy3 and Cy5 label reactions) together, transfer combined reactions to a Microcon 30 filter and add 400 μL TE buffer, pH 7.4.

8. Centrifuge samples at > 13, OOOg until ~25µL remaining (~ 20 min). Then add 200 µL TE buffer, pH 7.4, to samples and spin until almost dry, < 7 µL (~ 10 min). Recover labeled DNA samples by inverting Microcon filter into a new 1.5 mL collection tube and centrifuge at half max speed for 1 min.
9. Bring sample to 7 µL with TE buffer, pH 7.4, and transfer to a new 0.5-mL tube.

Hybridizations

To reduce the potential effects caused by variation in array quality, each hybridization should be performed at least twice. Because the chlamydial genome was printed twice per slide, there was a minimum of four data points for each locus.

1. Add the following to each sample: 0.67 µL tRNA 10 mg/mL, 1.90µl_ 20X SSC, 1.35 µL 2% (w/v) SDS.
2. Heat denature samples by incubating at 950C for 2 min. Perform a quick spin in centrifuge to pellet samples and let samples cool at room temperature for 2 min before applying to microarray.
3. Apply each sample to microarray and seal with rubber cement.
   a. The array is invisible after postprocessing. Therefore, to ensure proper placement of a cover slip, make a template of your array using a clean glass slide, or by drawing on a piece of paper. Be sure to indicate the corners of the array and where the cover slip should lie.
   b. Lay each microarray slide on top of the template and, just inside the marked cover slip corners, apply a small microdot of rubber cement with a needle and syringe to each corner.
   c. Apply the ~10µL sample to the microarray slide. Place an appropriate size cover slip with bent precision forceps right on to of the four small dots of rubber cement. Try to free any air bubbles that might be present under the cover slip by gently tapping on the top of the cover slip with the forceps.
   d. Completely seal the edges of the cover slip with the rubber cement. To avoid sticking of the cover slips to the top of the hybridization chambers, let the rubber cement dry before sealing the hybridization chambers.
   e. Place the prepared and sealed microarray glass slide into a chamber compartment in a GeneMachines HyChamber and place ~100 µL of TE buffer, pH 7.4, in each chamber compartment.
   f. Seal each GeneMachines HyChamber completely and place in appropriate temperature water bath for 12 h to overnight. (All chlamydial hybridizations took place under a glass cover slip in a 75 °C water bath overnight except for the MoPn vs DIUW-3 comparison, which transpired in a 65 °C water bath overnight.).
4. Take microarray glass slide out of the HyChamber and immediately submerge the slide into a flat container filled with Wash Solution 1.
5. Using a clean razorblade, remove the cover slip by lifting it at the corner and pulling up.

Transfer the array to a slide rack submerged into a container filled with the Wash Solution 1 and plunge rack up and down for 2 min to wash slide.

6. Place the rack holding the array slides into a new container filled with Wash Solution 2 and plunge up and down for 2 min.
7. Place the rack holding the array slides into a new container filled with fresh Wash Solution 2 and again plunge up and down for 2 min.
8. To dry the slides, place rack holding array slides in table top centrifuge and spin for 5 min at 50g.
9. Scan each microarray glass slide using a GENEPIX Scanner 4000A and analyze the resulting 16-bit TIFF images using GenePix Pro 4.0 (or equivalent) software.

Mathematical Calculation

For each isolate, a normalization factor is calculated for the signal intensity of each wavelength and used to standardize any variation between replicate slides. The signal ratio of the normalized intensities between each wavelength is then determined for each locus.

Because there are several data points for each locus of an isolate, the average ratio and its standard error is calculated for each region. In the chlamydial microarray assay, the signal ratio between the intensities from the reference DNA (635 nm) and the test DNA (532 nm) was used; it is the level of departure from a 1: 1 ratio that indicates the degree of sequence polymorphism at a locus.

1. Normalize the signal intensity of each wavelength for each isolate to one chosen slide of that isolate.
   a. For each isolate, the normalization factor for a given wavelength for any slide Y is determined as the ratio of total signal intensity of a chosen slide X relative to the slide Y as shown next:

   $$NF_Y = \Sigma\ I_Y / \Sigma\ I_X$$

   Where NF_Y is the normalization factor for any slide of an isolate at a given wavelength, $\Sigma\ I_Y$ is the total intensity from any slide of an isolate, and $\Sigma\ I_x$ is the total intensity for the specific slide of the isolate chosen to be used as the basis for normalization.
   b. The normalized signal intensity of each locus on slide Y is then calculated by multiplying the signal intensity from each wavelength of that locus by its respective NF_Y for that slide.
2. For each slide of an isolate, calculate the ratio of the median signal intensity for each normalized locus on the array.
3. For each locus of an isolate, calculate the mean and standard error of the mean for the signal ratios determined in step 2.

Detection of Locus

To detect loci on the array that may lead to false-positive data, the assay is performed using identical test and reference DNA, both of which are the same as the DNA used to construct the slide. With an expectation that all loci should have equal ratios of hybridization of the labeled DNA, those that are significantly different are eliminated from future analyses.

1. Follow the steps from described elsewhere in this chapter. on replicates from hybridizations that used the same DNA for both the test and reference samples.
2. Determine which loci are statistical outliers (e.g., Z-test, 95% confidence interval).
3. Remove the data at these loci from all subsequent investigations.

Corelative Studies

If two or more genomic sequences are known (and one of them was used to build the array), then these can be used to demonstrate the correlation of the signal ratio and sequence identity between the loci of the two genomes. However, this step can be eliminated if two genomic sequences are not available as it has been established for several bacteria.

1. Follow the steps from described elsewhere in this chapter for replicates from hybridizations that used test and reference DNA from samples that have fully sequenced genomes.
2. Determine percent nucleotide sequence identity between the two genomes for each locus represented on the microarray.
3. Graph the percent identity described the log2 mean signal ratio for all the genes.

Assessment of Samples

By ranking the average ratio of the median from highest to lowest for each sample strain, assessments can be made about which regions are the most variable when compared to the reference strain. This also allows comparisons between strains regardless of total genomic identity.

1. Follow the steps from described elsewhere in this chapter for each sample.
2. Sort the average of the signal ratios for each locus from highest to lowest for each isolate, and rank them with one corresponding to the highest average.
3. Those ranked highest have the most sequence variability compared to the reference.
4. Compare variable loci among the isolates and with observed attributes.

POINTS TO REMEMBER

1. Prepare this solution fresh each time.
2. Higher temperatures were used to increase binding specificity; lowering the salt concentration or adding formamide may also help.
3. The value of the signal intensity used for each wavelength is the background subtracted from the median intensity.
4. The numerator should be the wavelength that corresponds to the reference strain; this will give ratios one and above.
5. If the standard error of the mean is large for a locus, look at the individual data points and their respective spots in the raw TIFF image file as there may have been a problem

with the slide at that region. If there was something that artificially altered the data, that point should be eliminated from analysis.

6. For the chlamydial array, strain D/UW-3 was used as both the test and reference strain.
7. Although an unknown degree of bias may be introduced at these loci, a very high signal ratio (>5) in a test isolate may still be indicative of a gene deletion or high degree of polymorphism.
8. For the chlamydial array, strains D/UW-3 and MoPn were used.
9. Be sure to analyze only the sequences of the PCR products used to make the micro array.
10. A large mean signal ratio may be the result of a full or partial deletion; further analysis by other methods, such as PCR or Southern blot, would be necessary for confirmation.
11. Sequencing a few loci that were found to be either variable or highly similar for verification purposes would add confidence to all the findings in the study.

Chapter 5

Genetic Mapping

In contrast with genetic maps, physical maps of genomes or genome segments (e.g., chromosomal DNA) relate genome positions to each other using *physical distances* measured along the DNA helix axis. Distances between positions are often expressed in base pairs (bp) or multiples thereof (for example, *kilobases* (kb)-bp x 1000). Large-scale physical maps are useful for some genome *sequencing strategies*, and they are even more important when studying *genomes* that have not been sequenced.

However, sequencing "*factories*" can now churn out so much sequence per day that for small genomes it is easier and faster to determine the genome DNA sequence and identify the restriction sites computationally. *Markers* on the physical map allow investigators to retrieve particular regions of interest for further study. *Restriction* endonuclease (or restriction enzyme) cleavage sites represent one type of physical marker. *Genetic markers* (defined by mutations in genes) may affect the sizes of restriction fragments, leading to a *correspondence* between a genetic and a *physical marker*.

Examples are mutations that destroy a restriction site (a restriction site polymorphism), or deletions, insertions, or inversions. Other types of markers are portions of a gene sequence. The presence of such markers can be detected by *hybridization* reactions, even if the complete DNA sequence is unknown. Sequences that are not part of genes can be used as physical markers, and these include segments of DNA containing variable numbers of tandem repeats and *sequence-tagged sites,* or *STSs*.

The presence of a particular STS is revealed by the presence of PCR. reaction products obtained by use of a particular primer pair so chosen that they allow amplification of DNA from only one location in a particular genome. *Obviously*, the ultimate physical map is the DNA sequence. In elsewhere chapter in this book, we discussed the properties of *restriction* digest products within the context of eventually constructing a restriction map.

Recall that a restriction map is a display of positions on a DNA molecule that can be cleaved by one or more restriction *endonucleases* and that this is one particular example of a physical map of a genome. As we will see below, reconstructing a restriction map from the restriction digest products can be *computationally* very hard. Laboratory workers circumvent this *complexity* by a number of experimental techniques, including use of multiple restriction enzymes, end labeling of DNA prior to digestion, analysis of *incompletely* digested end-labeled products, and *hybridization*.

They also may employ genetic variants of the molecules being mapped (ones containing insertions, *deletions*, or *inversions*). Laboratory workers usually perform map construction by using a *massively* parallel, cranially mounted neural network (the brain).

Optical mapping is an experimental strategy that preserves fragment order after digestion, but it requires nonstandard laboratory equipment in addition to more *sophisticated statistical* tools to allow for fragment length measurement errors and products of incomplete digestion. DNA molecules can be too long to be handled conveniently without breakage. For this reason, *genomes* are intentionally fragmented into pieces of *appropriate* size for cloning into convenient cloning vectors.

For example, pieces cloned into lambda vectors may be approximately 20 kb in length, those cloned into *cosmids* will be approximately 40 kb in length, and those cloned into BACs may be in the range of 100-300 kb. The inserts in these vectors can be individually mapped (the large-map problem is broken down into a large number of more tractable small-map problems).

Then the genome map is built up by merging the maps of the constituent cloned inserts. One of the *experimental* issues is how many clones must be characterized before the genome map merges into a single *continuous* segment. Physical maps can be informative for comparisons among organisms. For example, most unicellular *eukaryotes* and all *multicellular* animals and plants have organelles called mitochondria, which contain circular DNA molecules.

The orders of the approximately 37 genes in the mitochondria of various animals have been determined, and they are not identical. A model to describe differences in order is based on breakage/rejoining or reversal of DNA segments, leading to *permutations* of the gene order. It should be possible to model this process in a manner that predicts animal phylogeny (*branching* pattern connecting *lineages* of organisms to common ancestors).

This also has relevance when comparing bacterial strains with each other. For example, there are strains of *Escherichia coli* that are nonpathogenic and other' that are pathogenic. There are species of *Yersinia* that give rise to epidemics of plague and others that do not. By comparing genomes (including gene orders), it is possible to reconstruct steps in the *evolutionary* history of these organisms that may provide insights into mechanisms of *pathogenicity*.

PROBLEM IN COMPUTATIONAL BIOLOGY

This is an older, classical problem in computational biology, but we nevertheless describe it here to reinforce previous concepts and to provide experience with *computationally* complex problems. We will work with an idealized version of the problem that still illustrates the computational challenges.

Given the sizes of restriction fragments from digestion of a DNA molecule by restriction endonucleases A, B, and a combination of those enzymes, A + B, reconstruct the order of restriction sites in the undigested molecule:

Products generated by digestion with A, sorted by increasing size,

$= \{a_1, a_2, \ldots, a_n\}$;

Products generated by digestion with B, sorted by increasing size,
$= \{b_1, b_2, \ldots, b_m\}$;

Products generated by digestion with A + B, sorted by increasing size,
$= \{c_1, c_2, \ldots, c_{m+n-1}\}$.

The solution is a list of positions for cleavage by A or B such that the fragment set $\{c_1, c_2, \ldots, c_{m+n-1}\}$ is generated.

Data Collection

A sample of DNA is "digested" with a restriction enzyme until all sites on every molecule are cut. Agarose gel electrophoresis gives a way to separate DNA fragments (which are negatively charged) by their lengths. The molecules migrate through the porous agarose a distance proportional to the negative *logarithm* of their lengths; big pieces hardly move, and small pieces fall like stones. This principle also allows DNA to be sequenced, a topic we treat later.

The approximate fragment lengths are determined and listed in order of size. The problem is to reproduce the order of the pieces in the original DNA. That is, gel *electrophoresis* gives us the set of lengths unordered relative to the actual location in the DNA. For example, *digesting* phage lambda DNA (48,502 bp) with *EcoRI* gives the following lengths (in kb): 3.5, 4.9, 5.6, 5.8. 7.4, and 21.2. The correct order on the lambda genome is 21.2-4.9-5.6-7.4-5.8-3.5.

If the molecule has been sequenced, we can also learn this ordering by downloading the sequence from a database and searching the string for occurrences of GAATTC. How are restriction maps of *uncharacterized* DNA molecules determined experimentally?

Double Digests Problem (DDP)

As we indicated above, one experimental approach to the restriction mapping problem is to select another enzyme, digest with that enzyme, and finally digest with a combination of both enzymes. Inferring the two-enzyme map of the DNA from the resulting data is called the *double-digest problem (DDP)*. Here is a simple example for a linear molecule digested with enzymes A and B (in these examples, *fragment* sizes are taken to be integers):

Products produced from digestion with A : {2, 4};

Products produced from digestion with B : {1,5};

Products produced from digestion with A + B : {1, 1, 4}.

This "toy" problem can be solved by inspection. The "ends" in the double digest must be 1 and 4. This is because the shorter of the fragments produced at each end by *digestion* with A or B must lie wholly within the larger of the end fragments produced by A or B. This observation solves the problem. The solution is

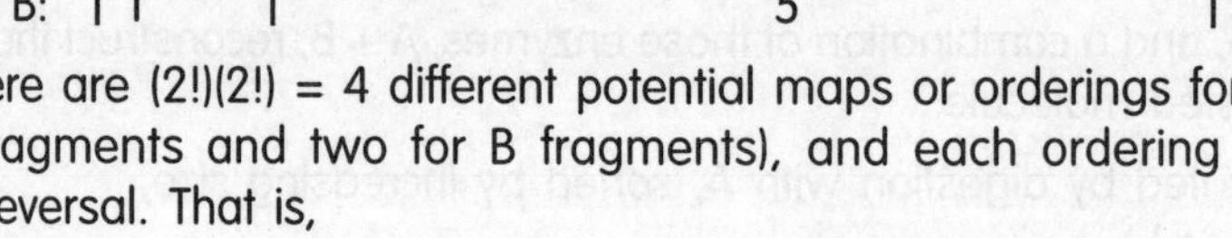

In general, there are (2!)(2!) = 4 different potential maps or orderings for this problem (two orderings for A fragments and two for B fragments), and each ordering has a *biologically indistinguishable* reversal. That is,

A: 2 4	and	A: 4 2	
B: 1 5		B: 5 1	

are reversals of each other. They correspond to reading fragment orders from one end of the molecule or the other.

It is easy for these problems to become more complex:

Products produced from digestion with A : {3, 4, 5}:

Products produced from digestion with B : {2, 4, 6};

Products produced from digestion with A + B : {1, 1, 2, 3, 5}.

Obviously, 5 must be at the end, and therefore 6 must be also. This is because there must be a fragment at each end from the A or B digests that is not cleaved after double digestion. Fragment 5 must lie within 6 since 6 is the only B fragment large enough to contain it. Since in this problem only integer values are allowed, one end of 5 must match one end of 6, which means that 5 is an end fragment from the A digest and 6 is an end fragment from the B digest:

(two fragments) -5

(two fragments)-6

Having determined the two end fragments, we have only four possibilities left to consider:

A:	345		345		435		435
B:	246	or	426	or	246	or	426

The last one is eliminated because it would produce too few double-digest pieces. The techniques used in these two examples obviously depend on the small sizes of the problems.

A more systematic method to predict the double-digest products is as follows: (1) take the fragment lengths, (2) produce maps of *positions, (3)* interleave the positions, and (4) take *successive* differences. For example, the digests above correspond to a molecule that is of length 12. The first map position is taken as 0, and the last position is 12. Starting at 0, successive positions on the map are produced by adding the lengths of *fragments*, starting from the left end:

Fragment ordering	→	Map positions	→	Interleaved positions
3 4 5		0 3 7 12		0 2 3 6 7 12
2 4 6		0 2 6 12		

Successive differences:

2—0, 3—2, 6—3, 7—6, 12—7 → 2, 1, 3, 1, 5

(This is a solution because the correct double-digest products are produced.) Now we try another ordering:

Fragment ordering	→	Map positions	→	Interleaved positions
3 4 5		0 3 7 12		0 3 4 6 7 12
4 2 6		0 4 6 12		

Successive differences:

3—0, 4—3, 6—4, 7—6, 12—7 → 3, 1, 2, 1, 5

(This is another solution!)

Now try the putative map positions of one of the other possibilities:

Fragment ordering	→	Map positions	→	Interleaved positions
4 3 5		0 4 7 12		0 2 4 6 7 12
2 4 6		0 2 6 12		

Successive differences:

2—0, 4—2, 6—4, 7—6, 12—7 → 2, 2, 2, 1, 5

This does not correspond to the observed products of double digestion. so this pair of *fragment* orders is rejected. The remaining possible ordering also fails to agree with the double-digest data, as you can verify for yourself.

The solutions to the double-digest problem are:

A: 3 4 5 and its reversal 5 4 3

B: 2 4 6 6 4 2

and

A: 3 4 5 and its reversal 5 4 3

B: 4 2 6 6 2 4

Finding the Data

Basically, we decided to try all orderings of each digestion product and to find out whether the predicted double digest fits the data. Here is an outline of the *method.* We assume that no cut site for A coincides with a cut site for B.

(1) Input fragment sizes:

$a_1, a_2, \ldots, a_n$ ($n-1$ cuts),

$b_1, b_2, \ldots, b_m$ ($m-1$ cuts),

$c_1, c_2, \ \ G_{m+n-1}$ ($m+n-2$ cuts).

(2) Produce position maps: .

$0, a_1, a_1+a_2, \ldots, a_1+a_2+\ldots+a_n=L,$

$0, b_1, b_1+b_2, \ldots, b_1+b_2+\ldots+b_{nm}=L.$

(3) Merge and put into order.

(4) Then take differences of adjacent numbers in (3).

(5) Check whether output of (4) is equal to $c_1, c_2, \ldots, c_{m+n-1}$.

When we solved our first tiny problems, we used special tricks for the data we had. Then we created a general approach to DDP that is correct and systematic.

What is the space efficiency? By this we mean memory commitments in our computer. For each map check, we needed space proportional to $n + m$. In addition, we need to save all correct permutations, and that we cannot estimate. The map check requires space proportional to $(n + m)$. Our examples were all small problems.

How complicated can things get with this approach? That is, what is the time efficiency? The number of possible orderings of the A fragments is $n!$, and the number of possible orderings of the B fragments is $m!$, and so there are $n! \times m!$ different combinations of position maps in (2) that require testing. As n and m become larger, the number of possibilities to test becomes huge!

n	1	2	4	6	8	10
$n!$	1	2	24	720	40,320	3,628,800

This means that a problem with ten A fragments and ten B fragments would yield over 10^{13} *subproblems* (2) through (4) to be *analyzed*. A very simple data set becomes *computationally* large very fast.

ALGORITHMIC SOLUTION

We now turn to specifying a method for solving computational problems such as DDP. The description of such a specification is called an algorithm. It should be kept in mind that this is a slightly higher level than computer code, although algorithms can and should result in code.

First of all, we need a set of basic operations such as addition, multiplication, etc. Next we describe some important aspects of algorithms.

(1) The input to and output from an algorithm must be precisely specified. (In the DDP, the input is three unordered lists (of integers), each of which must sum to the same number. The output is all orderings of the first two lists that solve the DDP.)

(2) The algorithm must consist of a well-defined series of operations.

(3) The algorithm must stop after a finite number of steps.

Finally, we must consider the notions of *correctness* and *efficiency.* Correctness is a mathematical issue of proving that the algorithm does what it should do. The DDP as we set it up takes a multiple of $n! \times m! \times (n + m)$ operations. This is an example of the time efficiency or time *complexity* of an *algorithm*.

The computational time complexity of an algorithm with input data of size n. is measured in "big *O*" notation and is written as $O(g(n))$ if the algorithm can be executed in time (or number of steps) less than or equal to $Cg(n)$ for some constant C. For example, adding up the numbers a_1, a_2, ... , a_n, can be accomplished in time $O(n)$ by the following method. Set $S = 0$. Then for $j = 1, 2....$ set $S = S + a_j$. On step n, S equals the desired sum. More generally, an algorithm has polynomial time complexity if $g(n)$ is a polynomial, and if $g(n) = a \times b^n$ for $a > 0$ and $b > 1$, the algorithm has exponential time complexity.

Our method for DDP is correct but hardly efficient. Can we find a more efficient method that is also correct, say one that takes at most a multiple of nm or even $n + m$ steps? The answer is very likely to be *no*. Here is the reason.

There is a famous problem called the *traveling salesman problem (TSP)*. Cities 1, 2, ... , *n* are located with various flights connecting them. The TSP is to find an ordering of cities $i_1 \rightarrow i_2 \rightarrow i_n \rightarrow i_1$ *so* that all cities are visited and the total route length is minimized. An example is shown below:

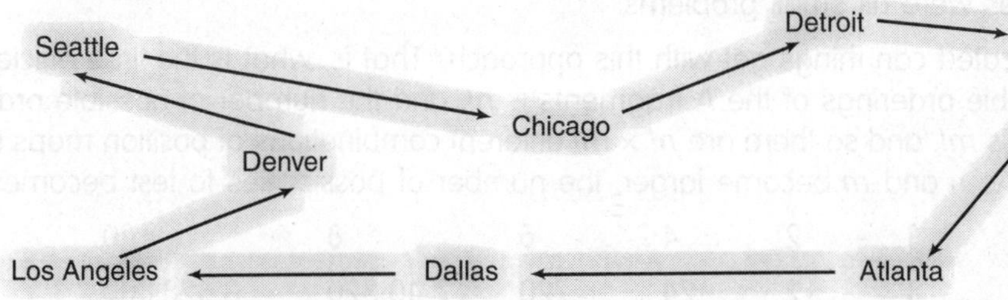

If the salesman starts in Seattle and follows the itinerary shown, is the trip shown shorter than if he were to end the trip by going from Dallas to Denver and then to Los Angeles and Seattle? This problem (in general) is one of the so-called NP-complete problems that have *equivalent* computational difficulty. It is unknown whether there is a polynomial method for solving these problems, but the *universal* belief is that there is no such method.

The DDP is one of these problems without a known polynomial algorithm. One more aspect of studying and *implementing* algorithms is space efficiency. As described above, our DDP algorithm has time *efficiency O(n! m! (n+ M))*. Space efficiency is measured in the same "big 0" notation.

We remark that in Section elsewhere in this chapter there was an algorithm for taking fragment permutations and producing the double digest lengths. That algorithm can be implemented in *0(n + m)* time and space. We have not described how to generate all *n! m!* pairs of permutations!

RESTRICTION MAPPING

When applying computational methods to biological problems, sometimes it is better to invest more effort in computational approaches, and other tunes it is easier to generate a different data set in the *laboratory*. We just showed that the double-digest problem rapidly becomes *extremely* complicated even for modest numbers of fragments.

Even a small microbial genome containing, for example, 1000 EcoRI fragments and 1000 *Hind*III fragments, if analyzed as described above, would require .analysis of an *astronomically* large number of subproblerns. This effort can be avoided by using a different experimental approach. In the examples above, molecules were digested to completion, so that all fragment orderings were completely disrupted.

There are two experimental approaches that retain ordering information. *Optical mapping* (Cai et al.. 1995) involves stretching individual DNA molecules onto a solid substrate, digesting them on the substrate, and then observing the ordered *digestion* products by fluorescence microscopy. Lengths of the product fragments (relative to included size standards) can be measured from recorded images in the *microscope* field, and lengths can also be determined from fluorescence intciisities.

This is particularly useful when employed with restriction endonucleases such as *NotI* that generate very large *fragments*. Another approach that preserves order information employs intentional incomplete digestion, as illustrated in Figure elsewhere in the chapter. At first consideration, this might seem to make analysis worse because if n fragments are produced by complete digestion, the total number of fragments produced by incomplete digestion of a linear *molecule* is

$$\binom{n+1}{2}$$

The experimental "trick" (Smith and Birnstiel, 1976) is to radioactively label (directly or indirectly) one end of the molecule. In Figure elsewhere in this chapter, the filled circle marks the labeled end fragment.

The mixture of completely and incompletely digested products can be resolved *electrophoretically*, and the sizes of the radioactive fragments can be measured (for example, after autoradiography of the dried gel or using a Southern blot, or with the help of a phosphorimager).

Scoring only the labeled fragments (only they appear on the autoradiogram) eliminates the

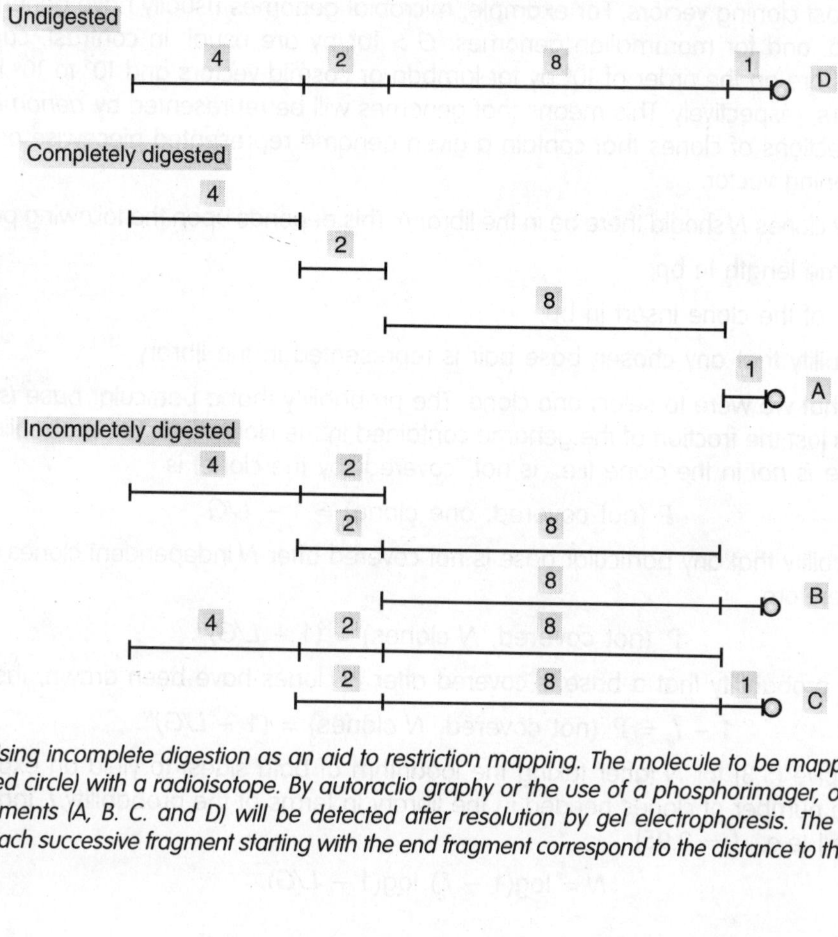

Figure 5.1: Using incomplete digestion as an aid to restriction mapping. The molecule to be mapped is labeled at one end (filled circle) with a radioisotope. By autoraclio graphy or the use of a phosphorimager, only radioactively labeled fragments (A, B. C. and D) will be detected after resolution by gel electrophoresis. The size increments separating each successive fragment starting with the end fragment correspond to the distance to the next restriction site.

background of incomplete digestion products that do not include the indicated end fragment. Now we list the labeled fragments in order of size:

0 1 9 11 15

This includes the map of restriction site positions measured from the right end!

In what was then an experimental tour de force, Kohara et al. (1987) applied this approach for *constructing* a restriction map of the *E. coli* genorne for several different restriction enzymes. They mapped clones in a *lambda* library and then *assembled* the larger map by overlap.

This general process, which in some form is used in most genome analyses, will be discussed in the next section. In the case of Kohara et al., individual lambda clones were indirectly end-labeled by hybridizing radioactive end probes to incomplete digests of cloned DNA that had been *transferred* to membranes by Southern blotting.

GENOMIC LIBRARIES

We indicated previously that analysis of genomes often proceeds by first breaking them into pieces. This is because genomes of free-living (nonviral) organisms are much larger than the sizes of inserts in most cloning vectors. For example, microbial genomes usually have genome sizes $G > 0.5 \times 10^6$ bp, and for mammalian genomes, $G > 10^9$ by are usual. In contrast, capacities of cloning vectors are on the order of 10^4 by for lambda or cosmid vectors and 10^5 to 10^6 by for BAC and YAC vectors, respectively. This means that genomes will be represented by *genomic libraries,* which are collections of clones that contain a given genome represented piecewise as inserts in a specified cloning vector.

How many clones *N* should there be in the library? This depends upon the following parameters:

G = genome length in bp,

L = length of the clone insert in bp,

f = probability that any chosen base pair is represented in the library.

Suppose that we were to select one clone. The probability that a particular base *is* recovered in that clone is just the fraction of the genome contained in the clone, L/G. The probability that any particular base *is not* in the clone (i.e., is not "covered" by the clone) is

$$\mathrm{P}\ (\text{not covered, one clone}) = 1 - L/G. \tag{5.1}$$

The probability that any particular base is not covered after *N* independent clones have been selected is therefore

$$\mathrm{P}\ (\text{not covered, } N \text{ clones}) = (1 - L/G)^N. \tag{5.2}$$

If f_c is the probability that a base is covered after *N* clones have been drawn, then

$$1 - f_c = \mathrm{P}\ (\text{not covered, } N \text{ clones}) = (1 - L/G)^N. \tag{5.3}$$

We can solve (5.3) for *N* (after taking the logarithm of both sides) to yield an expression for estimating the number of clones needed in the library in terms of the probability f_c that any base will be covered (e.g., $f_c = 0.95$):

$$N = \log(1 - f_c)\ \log(1 - L/G). \tag{5.4}$$

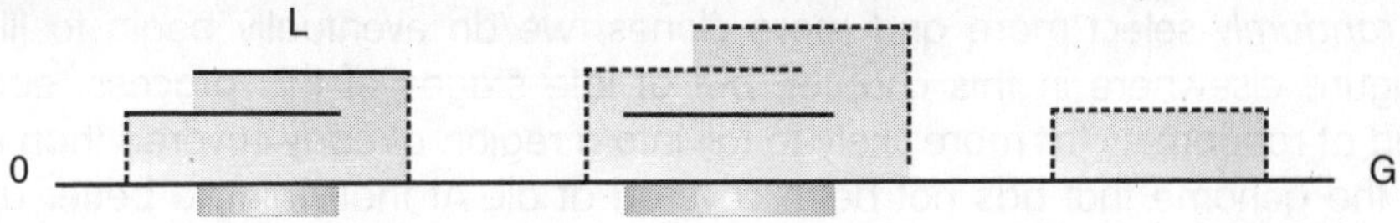

Figure 5.2: Coverage of a genome of length G represented by a genomic library, with inserts of length L for each clone. Some areas of the genome (shaded) may be represented multiple times in the library, whereas others are represented once or not at all. The coverage is the average number of times that each position in the genome is represented in the library. Broken lines indicate limits of the three contigs (gap-free assemblies of genome segments represented by two or more clones) shown.

Number of Cosmid Clones

How many cosmid clones are required in a library representing the *E. coli* genome such that $f_c = 0.95$?

We know that $G = 4.6 \times 10^6$ by and $L = 4 \times 10^4$ (for cosmids). Hence, *from* (5.4), $N = \log(1 - 0.95)/\log(1 - 4 \times 10^4/4.6 \times 10^6) = 343$ clones.

From Computational Example elsewhere in this chapter, we can calculate that the amount of DNA represented in the clones is 343 clones $\times 4 \times 10^4$ by/clone $= 14 \times 10^6$ bp. This is about three times the size of the *E. coli* genome. Some portions of the genome are represented multiple times and some not at all, but, on average, any position will be represented (covered) approximately three times in this particular library.

The *coverage* (the number of times, on average, that any base pair b is contained in inserts belonging to members of the library) is defined as

$$c = NL/G. \tag{5.5}$$

Equation (5.3) can be rewritten in terms of coverage:

$$1 - f = (1 - L/G)^N = (1 - (NL/G)/N)^N = (1 - c/N)^N \approx e^{-c} \tag{5.6}$$

Thus we find that

$$f_c = 1 - e^{-c} \tag{5.7}$$

The amount of coverage required to achieve different probabilities of including any particular base pair is

c	1	2	3	4	5
f_c	0.632	0.865	0.950	0.982	0.993

From this, we see that if we want to have a 99% chance of having a locus of interest in our library, we will need to clone five genome equivalents.

It is seen already that nearly half a million lambda clones would be required to represent *mammalian* genomes with a *reasonable* degree of *completeness,* and from the table above, we see that half again as many clones would be needed for f_f to be 0.99.

Actually handling libraries of this size (e.g., propagating clones, distributing clones, *arraying clones* on *hybridization* filters) is very labour-intensive, requiring robotics for *manipulating* the clones and databases for recording clone *inventory* and *annotation.*

As we *randomly* select more and more clones, we do eventually begin to fill in the gaps shown in Figure elsewhere in this chapter, but at late stages of this process, each successive clone, picked at random, is far more likely to fall into a region *already* covered than it is to fall into a region of the genome that has not been covered at all. At that point, a better approach is to switch to a directed experimental strategy for *gap closure-the* process of covering regions not previously covered.

Constructing Restriction Maps

As is the case for complete digestion with restriction enzymes, the cloning process disrupts the ordering of the cloned segments. The inserts are *generated* either by shearing the DNA or by restriction digestion, and usually the inserts are size-selected before ligation to the vector sequences. Size selection is needed particularly for vectors that accept inserts in a particular size range.

For example, lambda vectors accept inserts ranging from 9 kb to 20 kb, and cosmid vectors require inserts of length L ~ 40 kb. When individual clones are selected, their positions on the genome are unknown, and we must build up the complete restriction map from pieces of it. Developing a larger segment of the map from data for smaller segments is called the *"bottom-up"* approach. Using techniques such as the incomplete digestion method. the restriction map of individual clones can be determined.

If the inserts in two different clones contain some of the same restriction sites (i.e., generate restriction fragments of the same length), then the inserts may share a region in common (they overlap), and a larger, *contiguous* mapped region can be recognized by extending the physical map on both sides of the region of overlap. A *contig* is a genome segment represented by two or more overlapping clones.

A *contig* is part of a larger physical map, and its material representation is a particular set of clones. If the genomic library is very large, nearly all of the clonable portions (*heterochromatin* is difficult to clone) will be represented by clone inserts, and most regions will be represented many times on different clones, as shown in Figure elsewhere in this chapter. The desired outcome is to place each clone in its correct genome location, as shown in Figure elsewhere in this chapter.

A typical step in the process for achieving that outcome is illustrated in Figure elsewhere in this chapter. The *question* is, "How rapidly will the map approach completion as coverage c increases?" When should we stop *picking* clones at. random and move to directed gap closure? These questions can be answered using *Poisson statistics,* which were introduced in Chapter elsewhere in this book.

Assembly of Contigs

Figure elsewhere in this chapter illustrates the possible situations encountered for assembly of contigs. The process is dependent on recognizing overlaps, and it should be evident that some overlaps are too short to *recognize.* This could occur if the overlap is shorter than the average fragment size generated by the restriction endonuclease employed in mapping, for example. Also, because of the experimental error in *measuring* restriction fragment lengths, *identical fragments* might not be recognized as such.

Moreover, it is possible to generate similar-sized fragments from inserts from totally different parts of the genome. For this reason, it is usual to demand size matches among several fragments

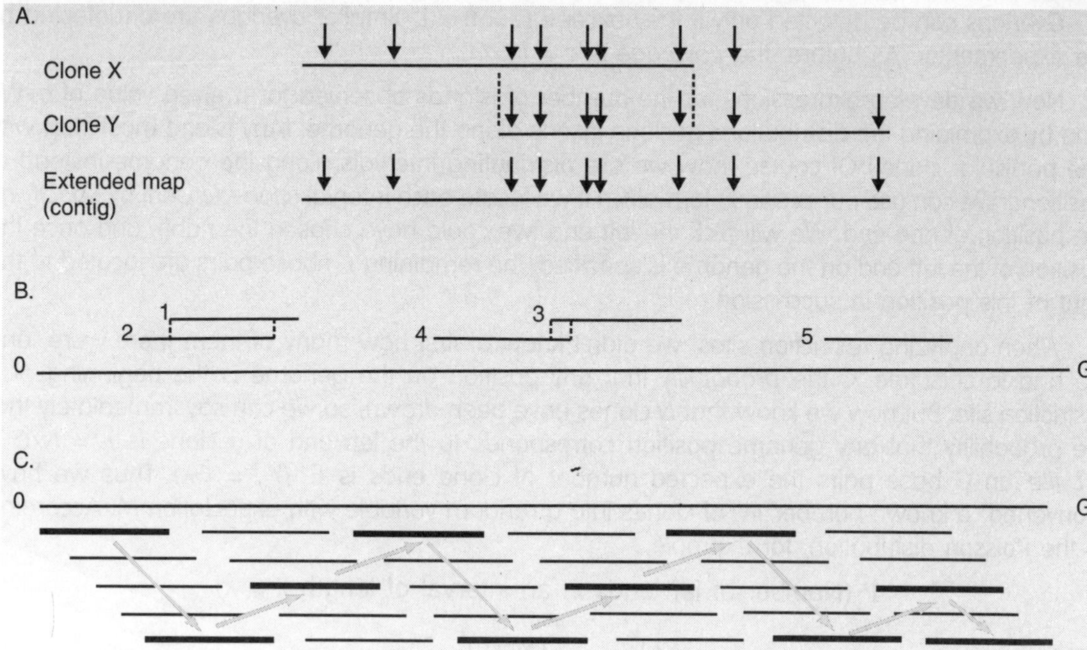

Figure 5.3: Mapping a large genome by assembling maps of representative clones. Panel A: Clones X and Y are seen to overlap because they share some of the same restriction fragments. This allows their maps to be joined to form a contig. Panel B: Cloned inserts can overlap sufficiently for that overlap to be detected (1 and 2), may overlap insufficiently for overlap detection (3 and 4), or may be singletons (5) having no context information supplied by neighbors. Contigs, singletons, and selected clones with undetected overlaps are called islands (see the text). Panel C: Once multiple clone overlaps have been constructed such that they span a whole genome or genome segment, there will be several clones covering each region. In this illustration, coverage c ~ 5. To reduce the redundancy, a subset of minimally overlapping clones is chosen (heavy lines) that still spans the whole genome. The path through the graph representing these clones (arrows) is called a Minimal tiling path.

from each clone before declaring that two clones overlap. In Figure elsewhere in this chapter, clones 1 and 2 represent a contig that can be recognized, while clones 3 and 4 represent a contig that cannot be recognized. Clone 5 is a singleton.

These three blocks of sequence (two contigs and a singleton) correspond to fixed points in the genome in a *"sea"* of genomic DNA. Contigs (those that are apparent and those not recognized) together with singletons constitute *"islands."*

To analyze progress in contig assembly, we list again the *parameters* pertaining to the problem (some of which are identical to the ones used in Section elsewhere in this chapter)

N = number of clones;

L = length of insert in each clone;

G = genome length;

Ω = minimum amount of overlap required for detection of that overlap;

θ = fraction of L corresponding to Ω; $\Omega = \theta L$.

Overlaps can be detected only if they exceed $\Omega = \theta \times L$. Smaller overlaps are undetected by the experimenter. As before, the *coverage* is $c = NL/G$.

Now we develop expressions for the number of islands observed for a given value of c. We start by examining the distributions of clone inserts along the genome. (Any island must start with one particular clone.) Of course, now we are distributing intervals along the genome instead of positions. We can use our previous formalism if we locate each interval (*cloned insert*) by *specifying* the position of one end. We will pick the left end (we could have chosen the right), and once the position of the left end on the genome is specified, the remaining *L-1* base pairs are located to the right of this position in succession.

When analyzing restriction sites, we didn't know at first how many of them there were, and we had to calculate λ, the *probability* that any position on the genome is the beginning of a restriction site. But now we know that *N* clones have been drawn, so we can say immediately that the probability that any genome position corresponds to the left end of a clone is $\lambda = N/G = c/L$ (*i.e.,* in G base pairs the expected number of clone ends is $\mathbb{E}(Y_N) = G\lambda$). Thus we have "converted" a known number *(N)* of clones into a random variable with expectation M. According to the Poisson distribution, for example,

$$\mathbb{P}\text{ (Number of left ends in an interval of length } x = k)$$

$$= \frac{(x\lambda)^k}{k!}e^{-x\lambda} = \frac{(xc/L)^k}{k!}e^{-xc/L}, \tag{5.8}$$

where *x* is the length of the interval along a DNA segment.

Now let's think about the number of apparent islands that will be observed. We have to use the word *"apparent"* because some clone pairs will actuilly overlap but not sufficiently to detect they will be counted is if they were two islands instead of one. We will enumerate the number of islands by counting their right ends. This is the same as the number of clones that are at the right end of an island. Let *F* be the expected number of islands. Then

$$\Gamma = N\mathbb{P}\text{ (A clone is at the right end of an island).} \tag{5.9}$$

Our problem is solved if we can calculate the probability that a clone is at the right end of an island. We can see what this probability is if we consider the diagram of a particular clone insert shown below (let's name it 8G11-the clone in row G, column 11 of microtitre plate 8):

8G11 ______ $(1 - \theta L)$ | θL

What is the probability that clone 8G11 is at the right end of an island? If there is another clone whose left end lies within θL of the right end of clone 8G 11. we can't tell whether there are any clones to the right of it. However. if the left end of a clone does lie within the segment of length $(1 - \theta)L$ diagrammed above, we can be sure that clone 8G11 is not at the right end of an island. The probability that the left end of another clone does not lie in the interval of length $(1 - \theta)L$ (*i.e.,* the probability that 8G11 is at the right end of a contig) is given by (4.8), with $k = 0$ and $x = (1 - \theta)L$. The expected number of islands then becomes

$$\Gamma = Ne^{-(1-\theta)LN/G} = Ne^{-(1-\theta)c} \tag{5.10}$$

To obtain an expression convenient for plotting, multiply and divide the expression on the right *by L/G* to obtain

$$\Gamma = (G/L)(Ce^{-(1-\theta)c}). \quad (5.11)$$

This result is the number of islands expected as a function of coverage and θ for given values of G and *L*.

The results for different values of θ are plotted in Figure elsewhere in this chapter. The verbal *explanation* for the shapes of the curves is as follows. When the mapping is started, there are no islands. At the beginning, as we draw each new clone for mapping, it is more likely to be a singleton (at that stage) than to overlap a clone already drawn, so the number of islands begins to climb. As the number of clones examined increases, existing contigs begin to increase in length as clones overlap their ends. So, later in the process, an ever-increasing proportion of the genome is being spanned by contigs.

Any new clones that fall inside an existing contig do not add to the number of contigs or islands. Some clones will overlap the left end of one contig and the right end of another, causing the contigs to merge. This means that the number of contigs and islands begins to drop later in the process. Eventually, the number of islands is expected to approach the final value of $\Gamma = 1$ (all

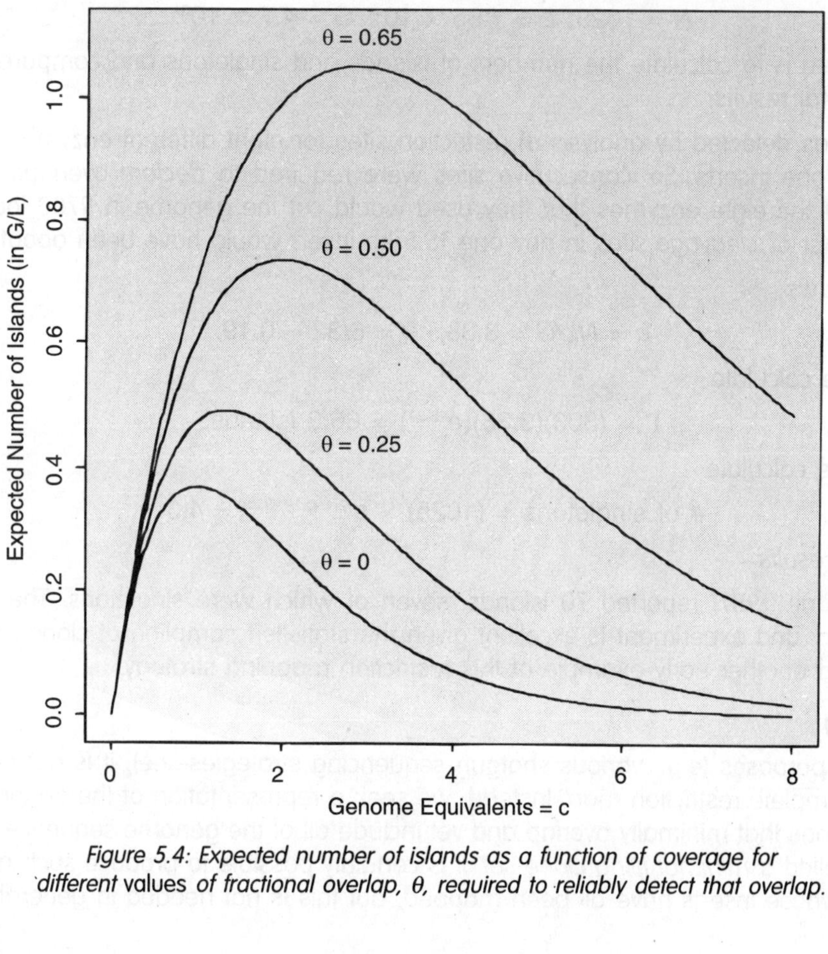

Figure 5.4: Expected number of islands as a function of coverage for different values of fractional overlap, θ, required to reliably detect that overlap.

clones merged into the single contig representing the entire genome). But the approach to Γ = 1 gets slower and slower as c increases. This is because, at the later stages of the process, most of the genome has been spanned, and the probability that a random clone will fall in the small gaps between the large contigs gets smaller and smaller.

We can estimate the number of *"singletons,"* clone inserts not overlapping others, by using the same argument. A clone insert C is a singleton if there is no other clone whose left end falls in the interval of length $(1 - \theta)L$ at the left end of C (probability = $e^{-(1-\theta)c}$ *and* there is no other clone whose right end falls in the interval of length $(1 - \theta)L$ to the right of C. This has the same probability as for the left end, so the joint probability that there is no other clone whose left end falls in the intervals $(1 - \theta)$ L to the left and to the right of C is $e^{-2(1-\theta)c}$ (the product of the two identical *probabilities*). The predicted number of *singletons* is then

$$\#\text{ of singletons} = Ne^{-2(1-\theta)c} \tag{5.12}$$

Reality check

Kohara et al. (1987) produced a complete restriction map for E. *coli* by using a random clone-mapping *strategy* employing inserts contained in lambda cloning vectors. The *parameters* were

$$N = 1025,\ L = 1.55 \times 10^4,\ G = 4.7 \times 10^6.$$

The problem is to calculate the numbers of islands and singletons and compare them with the experimental results.

Overlap was detected by analysis of restriction sites for eight different enzymes mapped in each of the clone inserts. Six consecutive sites were required to *declare* overlaps. Since they calculated that the eight enzymes that they used would cut the genome in 9700 locations, the average number of *cleavage* sites in any one 15.5 kb insert would have been about 32.

Calculated results

$$c = NL/G = 3.38;\quad \theta = 6/32 \approx 0.19.$$

From (5.11), we calculate

$$\Gamma = (303)(3.38)(e^{-2.84}) = 66.3 \text{ islands}.$$

From (5.12), we calculate

$$\#\text{ of singletons} = (1025) \times e^{-2(1-0.19)(3.38)} = 4.3.$$

Experimental results

Kohara et al. (1987) reported 70 islands, seven of which were singletons. The agreement between theory and experiment is excellent given the statistieil sampling of clones. Olson et al (1986) reported another early example of this restriction mapping strategy.

Fingerprinting

For some purposes (e.g., various shotgun sequencing strategies-see), it is not necessary to generate a complete restriction map. Instead, we seek a representation of the genome as a set of ordered clones that minimally overlap and yet include all of the genome sequence. Such a set of clones is called a *minimal tiling clone set.* It is certainly possible to produce such a set from a large library whose inserts have all been mapped, but this is not needed in general.

One way to produce a minimal tiling set is by fingerprinting the clones. A *fingerprint* of a clone insert is the set of fragments or fragment sizes produced from digestion by one or more restriction enzymes. Different inserts will produce different fingerprints.

As we indicated above, shared fragments from mapped inserts can be used to build contigs. With fingerprinting, the mapping step is skipped, and overlaps between two clones are declared when they share the requisite number of fragments whose sizes match within experimental error. With this procedure, restriction fragments are "*binned*" and partially ordered, even though the clones have not been completely mapped.

For example, suppose a digest of clones A, B, and C by the same restriction enzyme or combination of enzymes yields the fragment clone fingerprints indicated below. (Fragment sizes in kb are listed in order of size, and insert sizes are compatible with typical lambda cloning vectors. Assume that enzyme cleavages occur at the boundaries between the insert and vector sequences and that vector fragments have been removed from the list.)

A	B	C
7.0	6.0	6.0
5.0	4.0	5.0
3.5	3.0	4.0
3.0	2.5	3.0
2.0	2.0	2.0
1.0	1.5	1.0
	1.0	

Notice that A and C share fragments of sizes 5.0, 3.0, 2.0, and 1.0 kb. C and B share fragments of sizes 6.0, 4.0, 3.0, 2.0, and 1.0. If four matching fragments suffice to declare overlaps, then the following sets of fragment clusters are defined:

A : (7-0,3.5) (5.0, 3.0, 2.0, 1.0)

C : (5.0, 3.0, 2.0,1.0)(6.0, 4.0)

B : (6.0, 4.0, 3.0, 2.0, 1.0) (2.5, 1.5)

C : (5.0) (6.0, 4.0, 3.0, 2. 0, 1.0)

We don't know the order of the fragments enclosed in parentheses: we have arbitrarily listed them in order of decreasing size. But we have defined groups of fragments that occur as *neighbours* (in some undetermined order) in the restriction map. Since A and B both overlap C and share fragments 3.0, 2.0, and 1.0, we see that we can overlap all three clones. The particular clusters defined by each pair of clones allow us to further reduce the sizes of unmapped clusters:

A : (7.0, 3.5) 5.0 (3.0, 2.0,1.0)

C : 5.0 (3.0, 2.0, 1.0) (6.0, 4.0)

B : (3.0, 2.0, 1.0) (6.0, 4.0) (2.5, 1.5)

The result is a partial restriction map of the region represented on these clones, and this map was obtained by using the clone fingerprints *without mapping the individual clone inserts:*

(7.0, 3.5) 5.0 (3.0,2.0,1.0) (6.0, 4.0) (2.5, 1.5)

Given these clones and the mapping result, the region could be represented by clones A and B only. They would be the minimal tiling clone set for this region:

A : (7.0, 3.5) 5.0 (3.0, 2.0,1.0)

B : (3.0, 2.0, 1.0) (6.0, 4.0) (2.5, 1.5)

In this toy example, we had only three clones to compare. In actuality. t1ie·re is experimental error in determining fragment sizes, and the number of clones in the library may be very large. This could make the *fingerprinting* process tedious because there would be $N(N - 1)/2$ fingerprint comparisons to be made. If the library contained 100,000 clones, this would involve about 5×10^9 comparisons in all.

It is possible to reduce the labour of producing a minimal tiling clone set by prescreening the library for clones that overlap. We will describe this in more detail in elsewhere in this book, but we note here that it is relatively easy to determine the sequences for about 500 by into the insert from each end of the segment cloned. With this sequence, we can design PCR primers that will amplify DNA between them whereever that sequence is present and, in particular, within the inserts in overlapping clones.

We would then start with clone X and design primers for its "right" end. By using appropriate pooling schemes, it is relatively easy to identify other clones that overlap the right end of X because an amplified product can be produced from the appropriate (1oiw pools. Those are the only clones that need to be fingerprinted to deteruiiiie the ones that have minimal overlap with the right end of X.

If the left end of Y overlaps the right end of X, then we can design primers for the *right end* of Y, screen the library for clones that overlap its right end, fingerprint *those clones,* and continue the process until we have produced a minimal tiling path clone set. What we have just described is equivalent to employing *sequence-tagged connectors* to generate a minimal tiling clone set like those used for clone-by-clone shotgun sequencing.

Chapter 6

Gathering Genome Sequences

Like so many "*omics*" techniques, comparative genomics can mean different things to different people. Indeed, anyone who has performed a BLAST search for the sequence of their favorite protein using preexisting protein databases is carrying out *comparative* genomics in its most basic sense. The hope, in this example, is that other *researchers* have studied similar proteins and that their results, together with the *homology* found, can be used to *formulate* a *testable* hypothesis that will keep one busy for a few months.

In recent years, with increased availability of *bacterial* genome sequences, it has become possible to compare entire genomes, in order to make large-scale observations about evolution, *horizontal* gene transfer, and genes that might be responsible for *phenotypic* differences between the *comparator organisms*.

The aim remains the same as that of simple BLAST analysis, namely to formulate *hypotheses* that need to be tested. It is also possible to compare genome sequences without large-scale sequencing by using various *hybridization*-based approaches, where the aim is to locate DNA sequences found in only one of two comparator organisms.

This is proving a powerful and increasingly popular approach for locating genes involved in *phenotypes* specific to one particular isolate, but not shared with many *better-characterized*, or type, strains. *Comparative* genomics has now moved into the *functional genomics* era. In this incarnation, the comparison is of the expression of a gene or group of genes in different organisms having different *phenotypic* characteristics.

This review discusses previous and *ongoing* comparative genomic projects, and will describe *proteomic*, functional genomic, and hybridization-based approaches to comparative genomics. The *emphasis* will be on the analysis of bacteria relevant to human disease.

LARGE-SCALE GENOME SEQUENCES

Escherichia coli K12 is the most widely studied and most completely understood bacterium. It is used as a model system for other organisms, and the publication of its genome sequence in

1997 was one of the highlights of bacteriology research in the last century. Following publication of the genome sequence of a strain of *enterohemorrhagic E. coli* (EHEC) of serotype 0157:H7, comparative genomics was used to reveal some of the determinants responsible for *pathogenicity* in the latter (g. In fact, over 4 Mb of DNA sequence is shared by both strains, with very little *variation.*

This common DNA makes *E. coli* what it is, while the remaining DNA is the key to *strain-specific* phenotypes. Approximately 0.5 Mb is unique to K12, and around 1.5 Mb is unique to EHEC 0157:H7. Much of this unique DNA is located in strain-specific gene islands where the base composition (CG content) is not consistent with the background composition of the main part of the chromosome.

This reflects *horizontal* transfer of regions from donor species with a chromosomal base composition different from that of *E. coli.* EHEC 0157:H7 specific islands encode many known and candidate *virulence* factors, including *toxins, adhesion* and *invasion systems,* and extracellular antigenic determinants. These findings are being exploited to develop *sensitive* typing systems aimed at identifying virulent *E. coli* strains.

They have also fueled detailed analysis of the molecular basis of virulence in EHEC strains in general. Comparative genome sequence analysis *facilitates* the design of *multilocus* sequence *typing* (MLST) rationales. The comparison of K12 and 0157:H7 strains of *E. coli* led to an MLST scheme that was designed to undertake an evolutionary analysis of a *collection* of *pathogenic E. coli* strains. A mixture of 0157:H7 strain-specific genes and those shared by K12 and 0157:H7 were amplified using genomic DNA from members of the strain collection as *template,* and the *amplicons* were sequenced.

Phylogenetic analysis was then performed; the results allowed the evolution of the various strains tested to be predicted, and led to the conclusion that the same virulence factors had been independently acquired by a number of different *evolutionary* lineages of *E. coli* at different times. This remarkable finding highlighted the prime importance of a few specific virulence factors in the way *E. coli* isolates infect patients, and will form. the basis of further study.

Genomic Sequence Pseudomonas aeruginosa

In August 2000, the genome sequence of *Pseudomonas aeruginosa PAO*1 was reported and represented the largest bacterial genome to have been completely sequenced at that time. *Comparative* analysis of the products of individual open reading frames (ORFs) with those from other complete *genomes* using the amino acid sequence alignment program *blastp* showed that, at the time of the comparison, where homologs were present in other bacterial genome sequences, the *closest* matches were nearly always with *E. coli.*

Furthermore, gene order, average gene size, and average gene spacing in *P. aeruginosa* were remarkably similar to those of *E. coli.* The larger genome of *P. aeruginosa,* therefore, reflects greater *genetic complexity,* and not simply genetic rearrangements and spacing out of genes, or the occurrence of more recent gene duplication and/or *horizontal* gene transfer events.

This genetic complexity manifests itself as the presence of *significantly* more gene families than any other genome sequenced at that time, and not in simple duplications of gene families present in an ancestral bacillus. The conclusion drawn from these observations is that the diverse and unpredictable environment of *P. aeruginosa* has led to evolutionary pressure to generate new gene families that have *specialized* functions, but without losing *pre-existing* functions. It is far less likely that the opposite is true, i.e., relatively more *"specialist"* bacteria (like *E. coli)* have lost entire gene families; a few members of each would inevitably have been left behind by chance as a signature of their past existence. Understanding the forces that drive pseudomonads to evolve to cope with *complex* and varying environmental niches, is important in *controlling* the threat of these organisms in the clinical setting, where they are becoming *increasingly prominent.*

Genomic Sequence of Salmonella enterica

Members of the species *Salmonella enterica* represent prime targets for comparative genomic analysis. Isolates are generally very similar, with almost identical basic physiology, but different *serovars* can have very specific host ranges, and may cause infections characterized by very different *symptoms.* Hence, determining which serovar-specific genetic elements are present will *inevitably* further our understanding of these specific disease processes.

To date, however, only the genomes of strains of *S. enterica* sv. Typhi and *S. enterica sv.* Typhimurium %, have been published, together with a comparative analysis. *S. enterica* sv. Typhi is an important human pathogen, and is responsible for approximately half a million deaths worldwide per year. The genome sequence of the clinical isolate CT 18 was compared first with the genomes of *E. coli* strains K12 and 0157:H7, and subsequently with that of the laboratory strain of *S. enterica* sv Typhimurium LT2. *S. enterica* sv.

Typhimurium causes gastroenteritis, and has a wide host range, though LT2 is not particularly pathogenic One striking feature to have emerged from the comparison was that strain CT18 has around 200 genes that are shared with the three comparator organisms, but which contain mutations such that no functional protein product can be produced These mutated genes are called *pseudogenes.* There is no obvious decrease in the ability of CT 18 to survive outside the host compared to *S. enterica* sv.

Typhimurium LT2 and the *E. coli* strains, but this reduced functionality might be one reason why *S. enterica sv. Typhi* strains infect or colonize only humans, whereas *S. enterica* sv. Typhimurium and *E. coli* strains show a much broader host range. These pseudogenes have been found in all *S. enterica* sv. *Typhi* isolates examined to date, so they are not a unique property of strain CT18. Comparative analysis of CT18, LT2, and *E. coli* K12 also revealed the locations of previously characterized *pathogenicity* islands (PIs).

These genetic regions comprise the major differences between the two *S. enterica* serovars, with about 0.5 Mb of DNA being unique to each genome. These areas are undoubtedly responsible for host specificity and the differences in the types of disease caused by the two serovars, and will form the starting point for future work, particularly that aimed at the eradication of typhoid fever.

An additional benefit of the *S. enterica* sv. Typhi CT 18 genome-sequencing project was the complete sequencing of a 200 kb plasmid carried by this isolate. This plasmid is responsible for one of the most worrying features of modern *S. enterica* sv. Typhi strains, multidrug resistance, because it carries a large number of antibacterial, detergent, and heavy-metal-resistance determinants.

Indeed, this example highlights the potential of large-scale sequencing projects to determine a considerable amount of important information in the short time (and for a relatively small amount of money) it takes to sequence a large *plasmid crammed* with clinically relevant functions. Recently, Hansen-Wester and Hensel analyzed the PIs of published *S. enterica* and *E. coli* genome sequences in great detail. The rationale behind their analysis was that PIs tend to be located adjacent to tRNA gene loci that are anchors for bacteriophage *integration*, and so can be hot spots for the acquisition of genes by *horizontal* transfer.

There are around 80 tRNA loci in the *E. coli* K12 genome, and 2 kb on either side of each locus was aligned with 2 kb on either side of their homologs in the *E. coli* 0157:H7, *S. enterica* sv. Typhi CT18, and *S. enterica* sv. *Typhimurium* LT2 genomes. This sort of analysis is very time consuming, requiring considerable user intervention, and as such it is not usually included in proteomic sequence blastp analyses similar to those described for *S. enterica* and *E. coli*.

The analysis revealed four new regions of *horizontal* transfer specific to *S. enterica* strains. These regions have %GC contents that are far lower than the remainder of the *S. enterica* genome and contain a number of putative ORFs. Although *deletion* experiments did not reveal a role for these ORFs in the *progression* of infection in a mouse model, it is possible that they have a more subtle role that needs to be measured in a more *targeted* manner.

This is an excellent example of a focused comparative genomic project aimed at addressing a specific question, and represents the type of analysis that will be used increasingly in the future of comparative *genomics* as related to important human *pathogens*.

Genomic Sequence of Yersinia pesitis

The enomes of two *Yersinia pestis* strains, C092 and KIM, have been sequenced. Both are responsible for plague pandemics, although C092 is currently the most common strain. Its sequence appeared first, and has proved extremely important in our understanding of this highly virulent human pathogen. A large number of regions associated with horizontal gene transfer were located, together with regions showing the intriguing *phenomenon* of genome decay.

This occurs following evolution of an organism to grow in a new environment, via the acquisition of new genes by horizontal transfer. There is no longer selective pressure for the *maintenance* of genes that were important for its survival in its previous environment, and they are subsequently *degraded* by the *accumulation* of point and insertion mutations, until they appear as *pseudogenes*. This may be what is also happening to *S. enterica* sv. Typhi, albeit to a *lesser* extent.

The observation of gene decay in *Y. pestis* made by comparative genomics supports the previous proposal that *Y. pestis* evolved from an enteropathogenic *Yersinia pseudotuberculosis* isolate because, for the most part, it is the genes responsible for an enteric lifestyle that have decayed. When the *Y. pseudotuberculosis* genome is sequenced, it will presumably then be possible to target genes responsible for enteric infection for further analysis. Genome decay is the opposite of the situation in *P. aeruginosa*, whose niche is a fluctuating environment, where the addition of new functions is sometimes good, but the loss of functions is mostly bad.

A comparative genomic analysis of the *Y. pestis* KIM and C092 sequences revealed that, while there is a large degree of conservation between the sequences of specific genes, their location in the genome is dramatically different. The reason for this is not clear, but the observation is at odds with comparisons of other members of the Enterobacteriaceae, where there is a remarkably consistent gene order (synteny).

For example, the gene order is very similar in *E. coli* K12, *E. coli* 0157:H7, *S. enterica* sv Typhimurium LT2, and *S. enterica* sv Typhi CT 18, even though these organisms diverged over 100 million years ago. This is in contrast to the two *Y. pestis* strains, which are far more closely related in time. Putative virulence factor genes that differ between the two *Y. pestis* strains were observed following the comparative analysis, but the fact that these organisms are similarly pathogenic implies that these sequence differences are not particularly significant.

Indeed this is an example where comparative genomics was not entirely appropriate, because the biology of the comparator organisms are so similar that little important information was obtained from the analysis. The observed lack of synteny between the isolates showed that the genome is structurally less stable than those of many other organisms, implying that *Yersinia* spp. are capable of rapidly altering their genomes to adapt to new environments if evolutionary pressure is applied. The potential threat from this property should be taken very seriously indeed.

Genomic Sequence of Staphylococcus aureus

Staphylococcus aureus is one of the most important nosocomial bacterial pathogens, and the variability of its virulence, antigenic determinants, and susceptibility to antibiotics makes it an ideal candidate for comparative genomics. A number of sequencing centers set out to complete the genomes of various *S. aureus* isolates, but the first to be published was that of methicillin-resistant (MRSA) isolate N315 from Japan. In the same manuscript, the genome of the vancomycin-resistant (VRSA) strain, Mu50, was also reported.

Both strains fall into clonetype II-A. The results showed that *S. aureus* is remarkably flexible in its ability to share genetic information with other organisms in the environment, including even eukaryotes. The number of transmissible elements present in the two genomes is large, and, despite close phenotypic similarity between the two organisms, there are strain-specific insertions, including relatively large amounts of extra DNA, in Mu50.

Only about 50% of the genome could be traced back to a common ancestor *Bacillus/ Staphylococcus*, which is less than might be predicted given the phylogenetic relatedness between the two groups as determined by rDNA sequencing. This is a consequence of large amounts of recent genetic movement in staphylococci. As with the other organisms, many of the mobile genetic elements in *S. aureus* encode important virulence determinants.

A large proportion of the virulence genes have been previously reported to be important for softtissue infections. In addition, however, virulence factors were also found in the Mu50 and N315 genome sequences that have been associated with infection of *skeletal* tissue in other *S. aureus* strains.

Because neither Mu50 nor N315 have been found in skeletal lesions, no one had looked for these specific virulence factors, but it appears from the genome sequences that these isolates have the potential to cause such *infections*. Some of the *enterotoxin* genes located in Mu50 and N315 are strain specific, but most of the virulence factors identified are virtually identical in the two *strains*. Again, this is an example of two strains that are too *phenotypically* similar to make *comparative* genomics particularly revealing.

The exception to this was their antibiotic resistance, where the two strains *displayed* markedly different properties. As such, one major reason for the N315/Mu50 comparative genome analysis was to determine, en masse, the sequence of *antibiotic* resistance *determinants* present in both *organisms*.

All of those located are present on *mobile genetic* elements; Mu50 carries one additional *transposon*, which accounts for its increased spectrum of resistance. However, no obvious single locus was identified that might account for *vancomycin* resistance in Mu50.

In a subsequent analysis, whole proteomic sequences from Mu50 were compared with those from N315 in an attempt to predict which genetic differences might lead to vancomycin resistance in Mu50. The rationale was that since vancomycin resistance can be selected from an MRSA population, it is most likely to be caused by mutation of an intrinsic determinant, or group of determinants, present in all MRSA isolates.

This analysis was the first proteomic sequence analysis to be carried out, and the first comparative genome project to address a specific phenotypic difference between two organisms, and as expected the results obtained were complex. The sequenced proteomes of Mu50 and N315 show significant differences in *intrinsic* gene products, as a large number of point *mutations* were observed. Importantly, none of these were in genes *encoding* proteins with a predicted role in cell wall biosynthesis.

Given that the phenotype of *vancomycin resistance* is associated with overproduction and decreased *crosslinking* of the cell wall, only *mutations* in ORFs encoding proteins involved in this aspect of *physiology* were thought to be significant. A series of *putative* mutations likely to produce loss of function in cell wall and associated intermediary metabolism in Mu50, mainly resulting from frameshifts, were identified, and will provide the basis for future experiments to determine the genetic basis for *vancomycin* resistance in Mu50.

However, subsequent analysis has revealed that some of the apparent mutations in Mu50 might be artifacts. At least one of the genes reported to be mutated has been checked by *polymerase* chain reaction (PCR) analysis and found to be intact. Unfortunately, it seems that the Mu50 genome sequence was not properly completed prior to publication, and contains a high error frequency, due to incomplete coverage in some areas of the genome. This problem highlights what can happen when genome sequencing projects become too competitive.

The conclusions of any comparative analysis aimed at addressing a specific question should be tempered with caution concerning the quality of the sequence. Furthermore, comparative proteomic sequence analysis is limited to mutations in ORFs leading to *alteration/disruption* in protein sequence. When addressing a specific *phenotypic* question, it is equally possible that mutations in *noncoding* DNA sequences (e.g., promoter sequences) might lead to changes in the expression of genes, leading to the *phenotypic* effect seen.

This requires a hugely more time-consuming entire genome comparison, or a comparison of gene expression using one of the methods described already. Recently, the genome sequence of the community-acquired *S. aureus* strain MW2 has been reported. A comparison of this sequence with that of *hospital-acquired* N315 revealed that the majority of the genetic material was identical, and in the same gene order, but that there are a number of genomic islands that contain strain-specific genetic elements known to transfer horizontally.

In some cases, these islands carry virulence genes, and there are far more in MW2 than in

N315, though there are also a number of genes encoding functions less obviously associated with *pathogenicity*. It seems that these genomic islands are hot spots for transfer and rearrangement of *genetic material*, and may contribute to the high degree of *variability* seen among *S. aureus* isolates.

It is possible that genomic island allotyping methods, which rely on determining the sequences of these *hypervariable* islands in groups of strains, will be a useful future method of determining the pathogenic potential of particular isolates. Interestingly, MW2 shows far less evidence of acquisition of resistance genes than N315, which is *multiresistant*, and MW2 contains relatively few *mobile* genetic elements. It might be that the *community* environment of MW2 is easier to survive in.

Clearly, the use of antibiotics in hospitals is more intense than in the community, providing a stronger selective pressure for the acquisition of antibiotic resistance genes. In contrast, MW2 (but not N315) carries a *bacteriocin* determinant, which is believed to be important for community *carriage*.

Bacteriocins kill competitor bacteria, thus clearing the way for colonization by organisms that produce them (and which are immune to their effects). In contrast, it is thought that N315 colonizes hospitalized patients because prior antibiotic therapy kills potential competitor bacteria. The first example of shotgun comparative genomics was performed with an *S. aureus* isolate. Herron et al. sequenced random clones from a gene library made from genomic DNA of an *S. aureus* isolate associated with bovine *mastitis*.

Approximately 10% of the genome was covered, and these pieces of sequence, scattered about the genome, were aligned with the complete genome sequence of *S. aureus* N315. Even this limited analysis revealed sequences that were part of genomic islands, according to background sequence, and were specific to the bovine strain. The ORFs present in these islands might encode proteins essential for *infection* of the bovine, rather than the human host.

The analysis also showed a number of examples of horizontal gene transfer from gram-negative organisms not seen in previous *S. aureus* genome sequencing projects. Accordingly, *shotgun genomic* comparisons might have considerable value. If an ORF is located in a gene library from a particular species but is not found in the published genome of a member of the same species, it might have a strain-specific function. This function can be predicted and the prediction tested.

Genomic Sequence of A Streptococci

Acute rheumatic fever (ARF) and resultant heart disease is associated with infection by some strains of group *A Streptococcus* (GAS), but not others Recently, the complete genome sequences of two GAS strains were reported and compared. Strain MGAS8232 (serotype M18) is associated with ARF and strain SF370 (serotype M1.) is not: Approximately 10% of the genome is unique to each strain, with the remainder being identical. The unique portions are spread around the genome in DNA islands, some of which (including at least five phage genomes in ARF strain MGAS8232) are clearly associated with recent *horizontal* gene transfer.

Within the phage-associated DNA are 10 ORFs predicted to encode putative ARF-associated *virulence* factors; these are now the targets of research aimed at determining what causes ARF, and what might be done to prevent it in the future. This is an excellent example of how targeted *comparative* genomics can yield important information about the specific disease *associations* of otherwise highly similar *organisms*.

Genomic Sequence of Mycobacterium *Spp.*

Tuberculosis (TB) is arguably the most important human infectious disease, with *Mycobacterium tuberculosis* killing approx 4 million people each year worldwide. Although the genome sequence of only one clinical isolate of *M. tuberculosis* (H37Rv) has been published so far, the results have fueled a large number of *comparative* analyses.

The only direct genome sequence comparison among *Mycobacterium* spp. is between *M. tuberculosis* H37Rv and an animal-associated *Mycobacterium leprae* isolate. While *M. leprae is* recognized as an important human pathogen, it has never been possible to culture human *M. leprae* isolates in the laboratory in such a way that genomic sequence analysis can be performed; however, isolates that *colonize* certain animals (but do not cause disease) can be grown, albeit poorly, in the *laboratory*.

M. leprae has an unusual genome for a bacterium, because only approx 50% codes for proteins, with the remainder comprising large numbers of pseudogenes and regions of DNA that may have once coded for proteins, but which have mutated beyond recognition, possibly because of the loss of an important DNA polymerase proofreading activity.

Furthermore, the genome of *M. leprae* is about 1.2 Mb smaller than that of other *Mycobacterium* spp., probably because of recombination events between distant randomly repetitive sequences. The overall result of mutation and deletion means that *M. leprae* has less than a third of the functional ORFs carried by *M. tuberculosis* H37Rv. This gene decay is predicted to impair severely the ability of *M. leprae* to grow, and may well explain the difficulty in culturing this bacterium in the laboratory.

Genomic Sequence of Helicobacter pylori

Helicobacter pylori causes a wide variety of different disease presentations, and it is possible, though not entirely clear, that different strains might have acquired different virulence factors to allow them to cause different symptoms. Comparison of the genome sequences of *H. pylori* strains 26695 and J99 suggests that there are around 50-100 genes unique to each strain.

This interstrain variation most probably reflects gene acquisition rather than loss, and this suggests that *H. pylori* has a genome that is in continual flux. Evolutionary analysis of *H. pylori* strains has been considerably strengthened by comparative microarray analysis.

DETECTING GENE CONTENTS

This is a *hybridization*-based approach to determining gene content, though the use of microarrays allows a large amount of data to be collected at once. In this technique, the presence or absence of genes can be determined without further experiments being performed. However, despite its increasing popularity, *comparative microarray* experiments have major *limitations*.

Firstly, to produce a complete microarray for any organism, the sequences of all the loci in the genome encoding ORFs need to be determined, in order to design PCR primers to generate amplicons to be spotted onto the microarray slide. Therefore, whole genome microarrays are usually constructed for strains whose genomes have been sequenced. Unfortunately, it is not entirely clear at present how representative of the natural population of a particular organism some genome sequence strains are.

Second, only genes present in the genome sequence strain and absent in the comparator strain can be identified. This is because the result of hybridizing labeled DNA from the second comparator organism to the amplicons on the microarray is, in its most accurate form, binary; i.e., spots either become *labeled*, or they do not. Hence if genes are not present in the array (i.e., the *genome sequence strain*) but are present in the comparator organism, they will be missed.

This can compound the problem of how representative genome sequence strains are. The third problem, which applies to all hybridization-based comparative approaches, is that a gene identified as being present in the comparator organism on the basis of *hybridization* on the array, may not actually be intact, and may not actually encode a *functional protein*, cap with pseudogenes, a single *nucleotide* change can result in premature termination of a *translation* product, and the loss of the *function encoded*.

Despite these problems, as more genome sequences become available and more arrays are produced, comparative microarray analysis is becoming increasingly valuable, and there have been several recent successes with this technique. Comparative microarray analysis has been used to study the evolution of *E. coli* strains using the 4300 ORF E. coli K12 microarray, which was developed following the release of the E. coli K12 genome sequence, originally for gene expression (*transcriptome*) *analysis*).

In the comparative analysis, labeled total genomic DNAs from five E. coli strains of known evolutionary relatedness were individually washed over the array, and the number of positive *hybridizations* recorded in each case; 3782 ORFs were conserved in all strains. Analysis of these data allowed the *authors* to plot a complex predicted *evolutionary* path between the strains and, accordingly, to propose an evolutionary origin for them. This information is not only of *academic* interest, as the rules discovered might be useful in predicting the future *evolutionary* potential of important *pathogens*.

Dziejman et al. recently published an analysis of the relatedness of *Vibrio cholerae* isolates responsible for epidemic and pandemic diseases. The *V. cholerae* genome sequence strain, El Tor 01 N16961, was responsible for the seventh cholera pandemic, beginning in 1.961. Prior to this time, however, El Tor 01 strains caused epidemic disease, and there are examples of El Tor 01 strains that do not cause disease at all.

In order to determine some of the genes responsible for the pandemic nature of N 16961, comparative microarrays were used to identify genes not present in epidemic El Tor 01 strains and/or environmental strains, but present in a N16961 microarray representing 3632 ORFs. Among other discoveries about specific strains, the authors found 22 genes unique to N16961. Eleven of these genes were present as a *pandemic-specific* genomic island 1, and 8 more were found on *pandemic-specific* genomic island 2.

The equivalent regions of the genomes of the other comparator strains were sequenced; this confirmed the absence of these specific gene clusters and revealed the exact points at which these mobile elements were inserted into the chromosome. Unfortunately, as is typical of genome-sequence analysis, the possible functions of the genes, and the specificity of these functions in *pandemic disease*, were not entirely obvious from the gene sequences and their homologies with other known proteins.

However, the findings will stimulate a targeted analysis of gene function where no targets had been identified previously. Microarray analysis has also been applied to studies of *Mycobacterium spp.* For example, Behr et al. used it to identify 14 regions that are absent from the avirulent *M. tuberculosis* strain BCG Pasteur, which is used as an attenuated vaccine, but that are present in the *M. tuberculosis* H37Rv (virulent) genome sequence strain. Most of these deletions were also found in other attenuated vaccine strains. A review of the explosion of *comparative* microarray analysis in mycobacterial research has recently been published.

PROKARYOTIC GENOMIC HYBRIDIZATION (COMPARATIVE APPROACH)

It is reasonable to suppose that phenotypic variation among isolates of a bacterial species must be the result of genetic differences. Such variations often reflect horizontal gene transfer. It is important, if we are to understand which genotypes are responsible for specific phenotypes, that we be able to analyze differences in genetic complement between bacteria *quickly* and *simply*. However, it is not feasible to sequence the entire genome of strains representing every variant.

A method that focuses specifically on DNA sequences that are unique to one comparator organism, and so reduces the amount of information that needs to be sifted through, has the *potential* to enhance greatly our understanding of what makes *bacteria* express particular traits. Comparative hybridization is a method by which one can determine which sections of DNA are present exclusively in one of two comparator bacteria, without having to sequence either strain's genome.

The first part of this technique involves the separation of small DNA fragments generated by (frequentcutting) restriction digestion of genomic DNA from the first *comparator* organism, using 2D DNA *electrophoresis*. In this *procedure*, molecules are separated by size in the first dimension and by base composition in the second.

The DNA from this gel is then blotted onto a nitrocellulose membrane and the membrane washed with a solution containing radiolabeled DNA from the second comparator organism. The blot is then stripped and reprobed with *radiolabeled* DNA from the first comparator organism, where all the separated fragments become labeled.

A comparison of the two autoradiographs using sophisticated image analysis software identifies the DNA fragments found only in the first comparator organism, because these do not *hybridize* with labeled DNA from the second comparator. In order to sequence these unique fragments, a second 2D gel is run in parallel, using identically *fragmented* genomic DNA from the first comparator organism, where each fragment has been radioactively labeled at its 3' end.

The gel is then autoradiographed to locate organism one-specific DNA fragments, which can be eluted from the gel, ligated into a cloning vector, and sequenced. In the pioneering use of this method, Malloff et al. compared two strains of *P. aeruginosa* and discovered that one of the strains carried an *antibiotic-resistance cassette* containing an aminoglycoside acetyltransferase

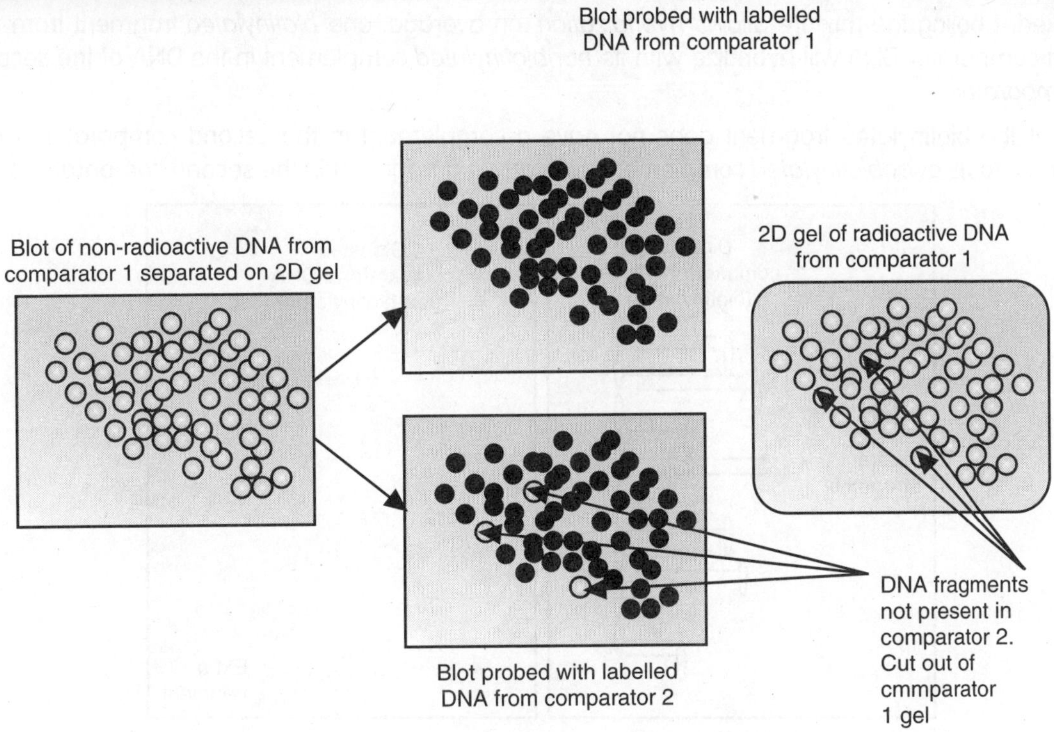

Figure 6.1: Method of detecting genomic hybridization.

gene. They highlighted one problem of the method, however, which is that many of the fragments can cluster in the center of the gel, so differences become impossible to resolve.

One possible way of improving the method is to produce a number of pairs of gels, each using DNA digested with a different *restriction* enzyme, thus increasing the chance that every part of the genome appears in an *uncluttered* part of the gel in at least one experiment. Another major problem with any hybridization-based approach to checking gene content is that it does not give information about the ability of the gene to be translated into a functional protein.

Pseudogenes (with frameshift mutations that stop the encoded protein from being expressed) still *hybridize* to their wild-type *comparator*, because there could be as few as one base difference in hundreds. To overcome these problems, comparative functional genomic approaches are required.

DELETING COMMON SEQUENCE

This technique is now widely used to find sequences unique to one of two comparator genomes. The principle is to remove all sequences common to both genomes from a mixed solution of genomic DNA. This is achieved by digesting both genomes with the same restriction enzyme, and *labeling* with biotin all the double-stranded fragments from the first comparator organism at their 3' ends.

The two sets of genomic fragments are heated to make the DNA single stranded, and then mixed. Cooling the mixture allows *hybridization*; on average, one *biotinylated* fragment from the first comparator DNA will hybridize with its *nonbiotinylated* complement in the DNA of the second *comparator*.

If the biotinylated fragment does not have a complement in the second comparator, it will anneal to its own *biotinylated* complement. Similarly, if a fragment in the second comparator does

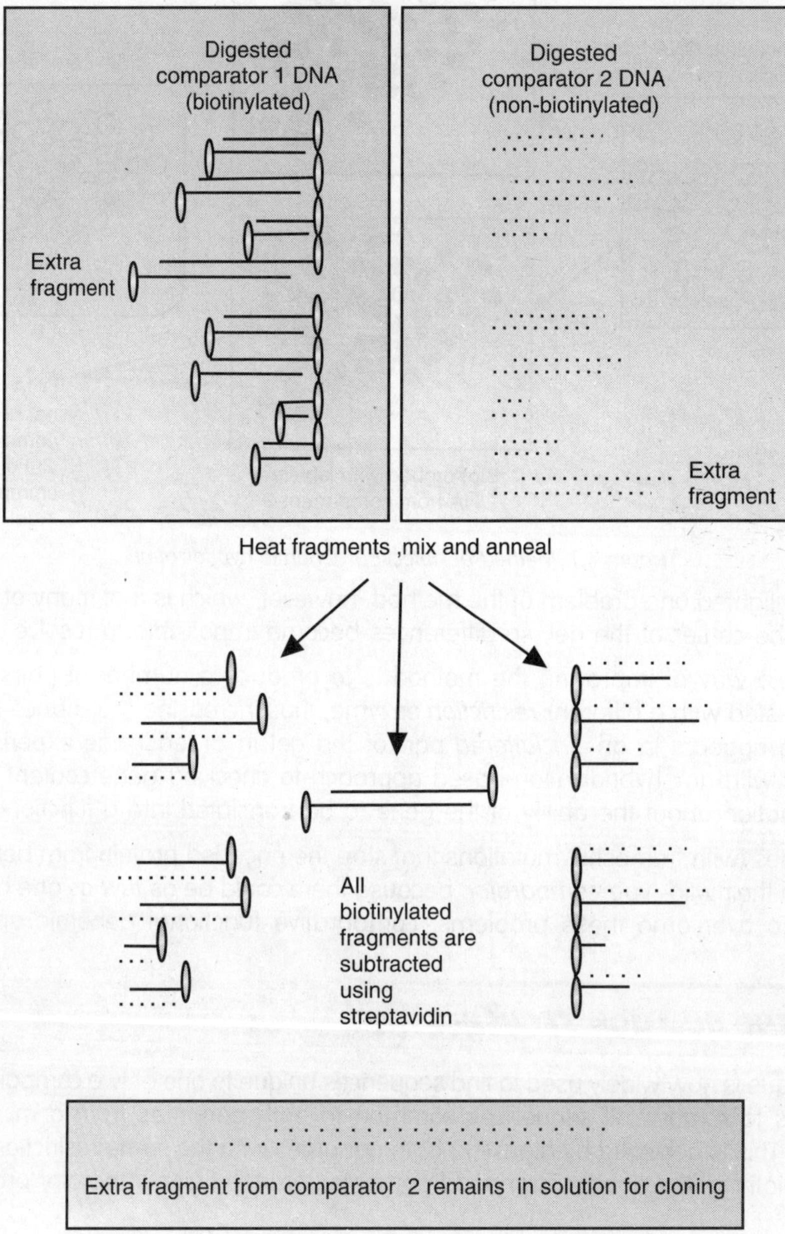

Figure 6.2. Method of deleting comman sequence.

not have a complement in the first genome, it will *hybridize* with its own *nonbiotinylated* complement. Only the latter group (DNA sequences exclusively found in the second comparator organism) remain completely nonbiotinylated.

All the biotinylated fragments can then be removed ("subtracted") by adding streptavidin-coated *Sepharose* beads or magnetic particles. Streptavidin binds to *biotin*, so when the beads are pelleted (by centrifugation or use of a magnetic bead *separator*), DNA fragments common to both comparators, or unique to the first, are pelleted. The fragments remaining in the *supernatant* can then be ligated to DNA linkers at both ends, and *amplified* using PCR with *primers* directed to the linker sequences.

The PCR products can then be *cloned* and *sequenced.* The whole procedure can be repeated following biotinylation of the second comparator DNA, in order to identify genes that are *uniquely* present in the first comparator organism. However, this approach is not suitable for detecting the presence of *pseudogenes.*

Methods very closely related to this are becoming popular for addressing the genetic reasons why one strain of a *bacterial* species is *pathogenic* while another is not. Brown and Curtiss used it to find unique genomic regions associated with an avian-pathogenic *E. coli* isolate, using the non-pathogenic *E. coli* strain K12 as the comparator. Pradel et al. used it to find genomic elements related to Shiga-toxin-producing *E. coli* strains.

The latter analysis used clinical isolates that do not produce Shiga toxin as comparators, and found some of the same genes that were previously observed in the 0157:H71 K12 genomic sequence comparison of Perna et al. confirming their *conclusion* that some *virulence* factors have been acquired by different lineages of *E. coli* at different times, *highlighting* their significance and the selective pressure for their *acquisition.*

It is important to undertake proper follow-up genetic experiments to confirm involvement of DNA sequences in virulence. For example, when true virulence genes are deleted or disrupted, the result should be an avirulent phenotype. Accordingly, this is another example of where comparative *genomics* allows *hypothesis* generation.

SUBTRACTIVE CLONING

In this method, genomic DNA from comparator organism 1 is digested with a blunt-end-cutting restriction enzyme, and the mixture of products is separated into two fractions. In fraction 1, the 5' end of each DNA fragment is ligated to a single-stranded linker sequence 1, to which a PCR primer can anneal. Fraction 2 is similarly modified, but the linker is an unrelated sequence 2.

The two tagged fractions are mixed together with a large excess of genomic DNA from the second comparator organism, which has been digested with the same restriction enzyme. The mixture is heated to make all *fragments* single stranded, and then cooled to allow hybridization. The *resultant* mixture is subjected to PCR with primers that target linker sequences 1 and 2. The only hybridization products that result in a PCR *amplicon* are those with linker 1 at one end and linker 2 at the other.

This occurs only when a DNA fragment from fraction 1 of the first comparator hybridizes with a complementary fragment in fraction 2 of the first comparator, but cannot happen if a similar complementary sequence is present in the second comparator, because the excess complementary DNA titrates out the linker-labeled DNA. Hence, only sequences found uniquely in the first comparator's genome are amplified, because no complementary sequence is present to titrate out the DNA.

The resultant amplicons can then be cloned and sequenced, and the method repeated to

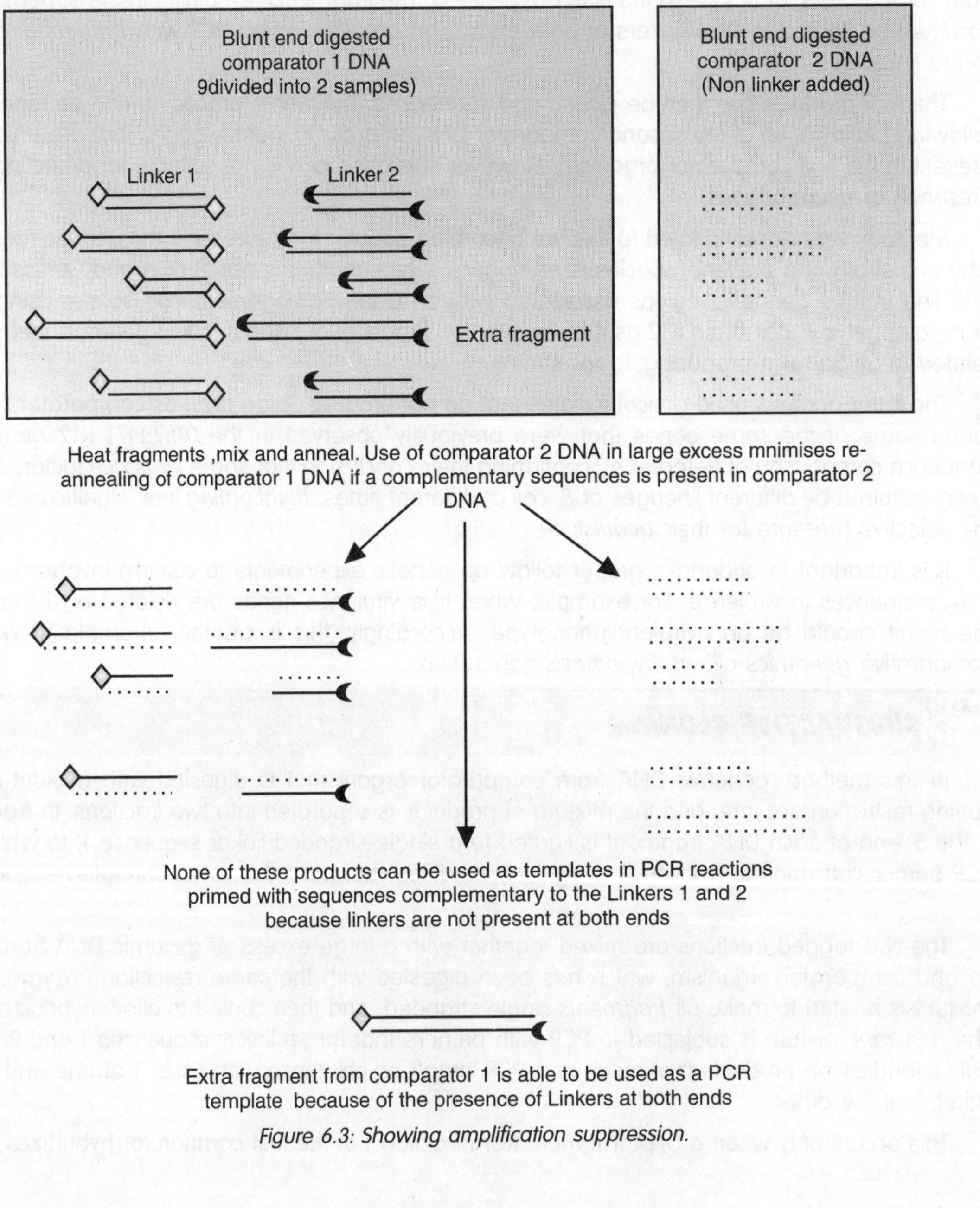

Figure 6.3: Showing amplification suppression.

find DNA unique to the second comparator organism. Subtractive cloning was used by Ahmed et al. to estimate the level of variation between *Campylobacter jejuni* isolates that colonized birds and humans. They successfully isolated 23 sequences found only in avian isolates, but could not rationalize a role for any of the sequences in pathogenesis.

Their analysis did, however, confirm the usefulness of this technique for determination of sequence variation in a group of organisms from one species, though the methodology is not entirely robust and suffers from a high false-positive rate, as do most hybridization-based methodologies. DNA hybridization is very sensitive to the *temperature* and salt concentration of the reaction buffer used, and reproducibility relies on careful *experimental* technique and good-quality equipment.

Bahrani-Mougeot et al. used *subtractive* cloning to determine DNA sequences found in the *uropathogenic E. coli* strain CFT073, but not present in *E. coli* K12 MG1655. Their analysis provided evidence of some virulence factors previously identified by Zhang et al. using a similar *methodology*, including fimbriae and *hemolysins*. They also identified some *potential* new virulence factors that need to be *characterized* further.

It is interesting to note that genome sequencing of *E. coli* CFT073 is currently under way, and preliminary sequence analysis has already validated the subtractive cloning experiments, adding evidence that this is a useful approach to comparative genomic analysis when a genome sequence is not available. With the commercialization of kits to undertake PCR subtractive cloning, this technique will undoubtedly become more commonplace. Again, it is not suitable for detecting the effect of *pseudogenes* on the complement of functional genes, and is likely to be most applicable to *determining* the *location* of genomic islands, which are very important in *encoding strain-specific* phenotypes.

ANALYSING COMPLETE GENOMIC

Analysis of complete bacterial genomes, and comparisons of two related strains either by sequence or hybridization analysis, is *increasingly* useful in determining what makes *comparator* strains *phenotypically* distinct.

However, there have been a number of problems with these approaches, as highlighted above. Moreover, they are generally too simplistic, as what functionally *determines* an organism's *phenotype* is its proteins, not its DNA. The activity of a protein is dictated both by its specific *amino acid* sequence and its level of *expression*.

The problem of pseudogenes-where the gene is present in both organisms, but no product is produced by one, potentially because of a single base change-has already been discussed. A single base change might also result in a specific amino acid change (e.g., in an enzyme's active site), which will lead to a *functionally* different protein product. These complexities can be addressed by sequence *proteome analysis* using blastp or CLUSTAL W.

However, we may also have to address the situation where an intact gene is present in two organisms, but its protein product is expressed in only one of them. This might be due to a mutation in the promoter, or in a transcription factor gene elsewhere on the chromosome. Analysis of the expression of genes, rather than just their sequence, is known as functional genomics.

Comparative Proteomics

In the functional genomics sense, *a proteome* is the complement of proteins expressed at any one time, while in the sequence sense, as used previously in this review, it is the complement of proteins with the potential to be expressed at some point in time. It is the former definition of the word proteome that will apply to this section. Clearly not all genes are *expressed* all the time, and subtle changes in gene expression from one *organism* to another might lead to *dramatic phenotypic* differences.

The method of proteomics is discussed elsewhere in this book, but, put simply, it is a way of separating as many of the proteins expressed by an organism as *possibleas* distinct spots on a gel or as peaks by liquid chromatography-and then identifying which protein each spot or peak corresponds to. As such, proteomic differences between strains are observed as spots/peaks that have different *intensities* in one organism as compared with another. The most dramatic result is when a spot/peak is entirely absent in one profile.

The protein is then recovered from the spot/peak in the other comparator profile, and identified by *proteolytic digestion* and mass *spectrometry*, or by N-terminal sequencing. Care is needed when comparing profiles from organisms whose genome sequences are not known, because point mutations in a gene can lead to amino acid changes in the protein it encodes, leading to altered electrophoretic and *chromatographic* properties, and the observation of one spot/peak *disappearing* and one appearing in the *comparator*.

Such false leads can result in a great deal of wasted time and resources. Difficulties with these techniques mean that they have been rarely used in *comparative* genomics (i.e., to compare different isolates). However, Betts et al. performed a comparative proteomic analysis of *M. tuberculosis* strain H37Rv, whose genome sequence has been published, and strain CDC 1551 (whose genome sequence has been determined, but not yet published).

These strains share > 99% sequence identity, but have significant phenotypic differences, potentially explained by changes in gene expression. The authors compared spots on 2D gels, and identified all the differential ?y expressed spots using mass spectrometry. Only 17 spot intensity differences were seen; the proteins representing 12 of them were identified. Four of the spot intensity differences were mobility variants caused by amino acid changes.

Three of the proteomic differences were found to represent proteins encoded on a DNA element found in only one of the comparator organism genome sequences, validating the *proteomic* approach used. Unfortunately, this proteomic analysis did not provide a reason for the *phenotypic* differences seen between the organisms, though the (as yet unpublished) *comparative* genomic analysis did highlight a number of blastp differences that might explain the *phenotypic* differences.

These important genetic changes were not reflected as proteomic changes; hence, either some of these genes were not expressed in the *comparator* encoding them under the conditions tested, or their respective spots were not resolved. Poor resolution is a *major problem* with 2D electrophoresis, though it is less problematic when analyzing organisms expressing relatively few proteins, such as *M. tube rculosis.*

Chapter 7

Mapping SNPs

The completion of the *human genome* project has allowed for the identification of millions of single nucleotide *polymorphisms* (SNP) loci, which makes them ideal *markers* for various genetic *analyzes*. Because of their abundance, even spacing, and stability across the genome, SNPs have significant advantages over RFLPs and *microsatellite* markers as a basis for *high-resolution* whole genome *allelotyping* with *accurate* copy number *measurements*.

High-density *oligonucleotide* arrays have recently been generated to support large-scale high throughput SNP analysis. It is now possible to genotype hundreds of thousands SNP markers using a single GeneChip *Human Mapping Array*. The current available *Affymetrix* SNP array platforms include GeneChip Mapping 10K, GeneChip Mapping 50K Array Xba 240, GeneChip Human Mapping 50K Array Hind 240, GeneChip Mapping 250K Nsp Array and GeneChip Mapping 250K Sty Array.

These platforms were initially designed for linkage analyses and casecontrol, family-based association studies, but because of its unique design, high *resolution*, and user-*friendliness*, *geneticists* have quickly adapted them for many different genetic applications, such as loss of heterozygosity (LOH) and copy number abnormality (CNA) analyses. In this chapter, we will provide a detailed *experimental protocol* of SNP array-based genomic *mapping assay* using GeneChip Mapping 50K Array Xba 240.

The experimental *procedure* described here can be easily adapted to the other Affymetrix SNP array *platforms*. The SNP array-based genotyping assay method devised for the Affymetrix GeneChip Mapping Assay is outlined in Figure elsewhere in this chapter. About 250 ng of genomic DNA is digested with a specific restriction enzyme (Xba I for the GeneChip Mapping 50K Array Xba 240 array-based assay) and ligated to a specific *adaptor* recognizing the cohesive overhangs created by *restriction* enzyme digestion.

All fragments resulting from restriction enzyme digestion, regardless of size, are substrates for adaptor ligation. A generic primer, which recognizes the adaptor sequence, is used to amplify the ligated DNA fragments using the PCR conditions that optimized to *preferentially* amplify fragments in the 250- to 1000-bp size range. The *amplified* DNA is then purified, fragmented to appropriate size (> 180 bp), labeled and *hybridized* to *GeneChip arrays*. The arrays are washed and stained on a GeneChip *fluidics* station and scanned on a GeneChip Scanner 3000.

The scanned array images are then analyzed using analytical software, such as GeneChip *Genotyping* Analysis Software (GTYPE) to reveal the allelic composition and produce the *genotype* calls for each SNP markers.

This procedure described here is adapted from the GeneChip Mapping 100K *Assay Manual*. The manual should also be *consulted* for any questions about the *protocol* (www.affymetrix.com).

REQUIREMENTS

Supplied by Affymetrix Affymetrix Reagents

This kit includes sufficient reagent for 30 arrays. The items included in the kit are listed below:

1. Adaptor, Xba: 38 µL of 5 µM annealed oligonucleotides, specific for ligation to Xba restriction site.

   5' A'ITATGAGCACGACAGACGCCTGATCT 3'

   3' AATACTCGTGCTGTCTGCGGACTAGAGATCT 5'

2. PCR Primer, 001: 1125 µL 10µM PCR primer, to amplify ligated genomic DNA.

   5' ATTATGAGCACGACAGACGCCTGATCT 3'

3. Reference Genomic DNA, 103: 25 µL of 50 ng/µL human genomic DNA (single source).
4. GeneChip Fragmentation Reagent (20 µL): DNase I enzyme, formulated to fragment purified PCR amplicons.
5. 10X Fragmentation Buffer: 225 µL of 10X Buffer for fragmentation reaction
6. GeneChip DNA Labeling Reagent: 60 µL of 7.5 mM Biotin-labeled reagent for end-labeling fragmented PCR amplicons.
7. Terminal deoxynucleotidyl transferase (TdT): 105 µL of 30 U/µL TdT enzyme used to end-label fragmented PCR amplicons with the GeneChip DNA Labeling Reagent (7.5 mM).
8. 5X TdT Buffer: 420 µL of 5X Buffer for labeling reaction.
9. Oligo Control Reagent (60 µL).

Equipment and Software

1. GeneChip Fluidics Station 450 (or GeneChip Fluidics Station 400/250).
2. GeneChip Hybridization oven.
3. GeneChip Scanner 3000 with high-resolution scanning patch.
4. Affymetrix GeneChip Operating Software (GCOS).
5. Affymetrix GTYPE.
6. GeneChip Human Mapping 50K Array Xba 240.

Hydrolytic Effect of Restriction Enzyme

1. Reduced ethylenediamine tetraacetic acid (EDTA) TE Buffer (TEKnova): 10 mM Tris-HCl, 0.1

mM EDTA, pH 8.0.

2. 20,000 U/mL *XbaI* (New England Biolab) also containing: I OX NE Buffer 2 and 100X: bovine serum albumin.
3. H_2O (BioWhittaker Molecular Applications/Cambrex): Molecular Biology Water.
4. 8-Tube strips, thin-wall (0.2 mL) and strip of 8 caps (MJ Research).
5. Thermal cycler.

Ligation

1. T4 DNA Ligase (New England Biolab), also containing T4 DNA Ligase Buffer.
2. Thermal cycler.

Amplification Technique

1. Platinum Pfx DNA Polymerase (Invitrogen Corporation): 2.5 U/µL, also containing lox Pfx Amplification Buffer, 10X PCRx Enhancer, and 50 mM $MgSO_4$.
2. 2.5 mM Each of dNTP (Takara).
3. 2 % TBE Gel (SeaKem Gold): BMA Reliant precast (2 %).
4. All Purpose Hi-Lo DNA Marker (Bionexus).
5. Thermal cycler.

Purification and Elution

1. Manifold - QIAvac multiwell unit (QIAGEN).
2. MinElute 96 OF PCR Purification Kit (QIAGEN).
3. Buffer EB (QIAGEN).
4. Vacuum Regulator to be used with QIAvac manifolds. The QIAGEN protocol requires ~800-mb vacuum.

Marking

1. 4 % TBE Gel (4 % NuSieve 3:1 Plus Agarose): BMA Reliant precast.
2. Thermal Cycler.

Hybridization

1. 5 M Tetramethyl ammonium chloride (Sigma).
2. MES hydrate SigmaUltra (Sigma).
3. MES sodium salt (Sigma).
4. DMSO (Sigma).
5. 0.5 M EDTA (Ambion).
6. Denhardt's solution (Sigma).
7. Herring sperm DNA (Promega).

8. Human Cot-1 (Invitrogen).
9. 20X SSPE (BioWhittaker Molecular Applications/Cambrex).
10. 10 % Tween-20 (Pierce): Surfactamps.
11. 1 mg/mL Streptavidin, R-phycoerythrin conjugate (Molecular Probes).
12. 0.5 mg Biotinylated Anti-Streptividin (Vector).

PROCEDURES

This procedure is adapted from the Affymetrix100K genotyping manual. The manual should also be consulted for any questions about the protocol.

Hydrolysis of XbaI

As this procedure processes DNA before PCR, *precautions* regarding *contamination* with template (human) DNA should be *stringently* observed.

1. Adjust the *concentration* of the target human genomic DNA to 50 ng/μL using reduced EDTA TE buffer (0.1 mM EDTA, 10 mM Tris-HCl, pH 8.0) in PCR *staging* room. See Note elsewhere in this chapter for the requirement of the target DNA.
2. Prepare *digestion* Master Mix (make 10 % extra): this step should be performed in a pre-PCR clean room or a PCR clean hood, where no template DNA is ever present. Make the mix as listed in Table elsewhere in this chapter. *Vortex* and *spin* briefly and keep on ice.

Table 7.1: *XbaI* Digestion Mixa[a]

Water	10.5 μL
NE Buffer	2 2 μL
10X BSA[b]	2μL
XbaI	0.5 μL
Total	15 μL

[a]Amounts for one reaction.
[b]BSA comes as 100X solution from New England Biolab. Diluted to 10X with H_2O before use.

3. Add 15 μL of the digestion Master Mix to 5 μL genomic DNA (50 ng/μL) in PCR tubes. *Vortex* and spin briefly.
4. Perform digestion in thermal cycler in the PCR *staging room* using following program: 37°C 2 h, 70°C 20 min, 4°C hold).
5. When the run is done, let it go down to 4°C and hold there for at least 2 min. Then, either proceed to *ligation* or store samples at -20°C.

Ligation

As this procedure processes DNA before PCR, precautions regarding *contamination* with template (human) DNA should be stringently observed.

1. Prepare the ligation Master Mix as listed in Table elsewhere in this chapter for number of samples that were digested plus 10% extra (to be performed in the pre-PCR clean room or a PCR clean hood, where no template DNA is ever present,)
2. Aliquot 5 µL of the ligation Master Mix to each tube that digestion was *performed* in (final volume in the PCR tube is 25 µL). This step should be performed in the PCR staging room.
3. Perform ligation reaction in the thermal cycler in the PCR *staging* room using following program: 16°C 2 h, 70°C 20 min, 4°C hold.
4. When finished, proceed to PCR *amplification* or store sample at –20°C.

Table 7.2: Ligation Master Mixa[a]

Water	10.5 µL
Water	0.625 µL
T4 DNA ligase buffer[b]	2.5 µL
Adaptor (*Xba* or *Hind*)	1.25 µL
T4 DNA ligase	0.625 µL
Total	5 µL

[a]Amounts for one reaction.

[b]Thaw T4 DNA ligase buffer on ice for at least 1-1.5 h.

If digestion was done on the same day, you can start thawing the buffer when you set up the digestion.

Amplification Technique

Three PCR amplification reactions are needed for each *ligation* product (sample) to produce sufficient product for array *hybridization.*

Preparation

1. Take the Pfx *polymerise* and PCR *Enhancer* out of the freezer and put on ice.
2. Vortex all of the reagents that were frozen and spin briefly to ensure they are well mixed.
3. Prepare PCR Master Mix as listed in Table elsewhere in this chapter for the number of *samples* that were digested three times, plus 5 % extra.

PCR Staging Room

1. Vortex the master mix on the *vortexer* in the hood to make sure it is well mixed. Store on ice.
2. Add 75 µL water to each tube in which *ligation* was performed. Mix by *pipetting* up and down several times.
3. For each ligation reaction (sample), aliquot 90 µL of the PCR Master Mix to three PCR tubes.

4. Aliquot 10 µL of diluted ligation into the appropriate PCR tubes. For the PCR negative control, use 10µL water.
5. Cover all tubes with strip caps, vortex, and spin briefly.

Mail Lab

1. Place the tubes in the PCR thermal cycler in the Main Lab. Recommended model MJ Tetrad PTC-225 96-well block (Bio-Rad), or ABI GeneAmp PCR System 9700-96 well block (Applied Biosystems).

Table 7.3: PCR Master Mixa[a].

Water	44 µL
PCR amplification buffer (10X)	10µL
PCR enhancer (10X)	10 µL
MgSO4 (50 mM)	2 µL
dNTP (2.5 mM each)	12 µL
PCR primer (10µM)	10µL
Pfx polymerase (2.5 U/RL)	2 µL
Total	90 µL

[a]Amounts for one reaction.

Table 7.4: PCR Programs for Amplification.

MJ thermal cycler			*ABI GeneAmp thermal cycler*		
Temperature	**Time**	**Cycles**	**Temperature**	**Time**	**Cycles**
94°C	3 min	1 X	94°C	3 min	I X
94°C	15s		94°C	30s	
60°C	30s	30X	60°C	45s	30X
68°C	60s		68°C	60s	
68°C	7 min	1X	68°C	7 min	1X
4°C	HOLD		4°C	HOLD	

2. Run following PCR program as listed in Table elsewhere in this chapter.
3. When the run is done, vortex the tubes, and spin briefly.
4. Check the PCR *amplification* product by run 3 µL of each PCR product on a 2 % TBE agarose gel, at 120V for approx 1 h. The PCR products should appear as smear on the gel, with average size between 250 and 2000 by.

Isolating the Amplification Product

1. Connect a *vacuum manifold* to a suitable *vacuum* source able to maintain ~800 mbar,

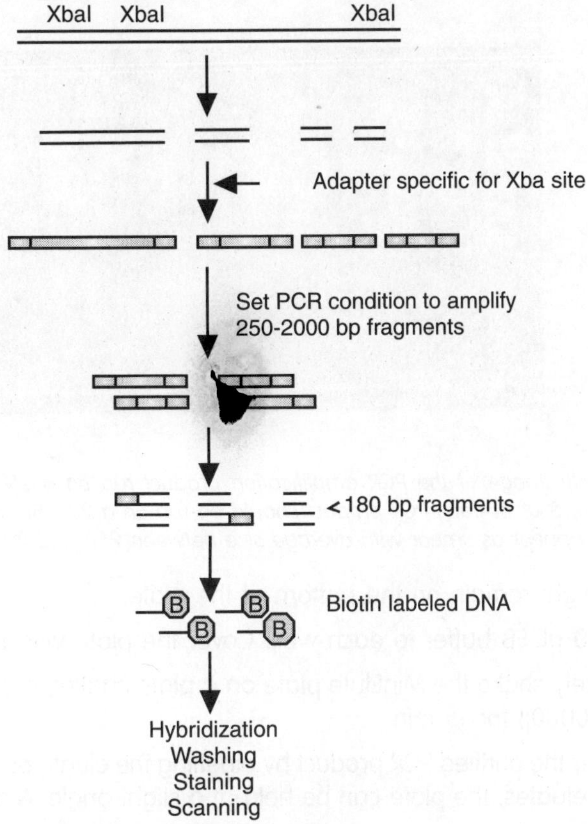

Figure 7.1: Outline of the GeneChip mapping assay: XbaI.

e.g., QIAvac 96 or QIAvac Multiwell Unit (QIAGEN). Place a waste tray inside the base of the *manifold.*

2. Place a MinElute 96 UF PCR Purification Plate on top of the manifold. Cover the entire plate with the aluminum cover. Cut off a strip for the wells you will use. Note: Mark the row with a marker so that we know it has been used.
3. Turn on the vacuum. Adjust the regulator to 800-900 mbar.
4. Consolidate the PCR product for each sample (three PCR reactions) into one well of the MinElute plate.
5. Maintain the ~800 mbar vacuum until the wells are completely dry.
6. Wash the PCR products by adding 50 µL *molecular biology* water and dry the wells completely (taking about 20 min). Repeat this step two more times for a total of three washes.
7. Switch off vacuum source and release the vacuum.
8. Carefully remove the MinElute plate from the vacuum manifold.
9. Gently tap the MinElute plate on a stack of clean *absorbent* paper to remove any liquid

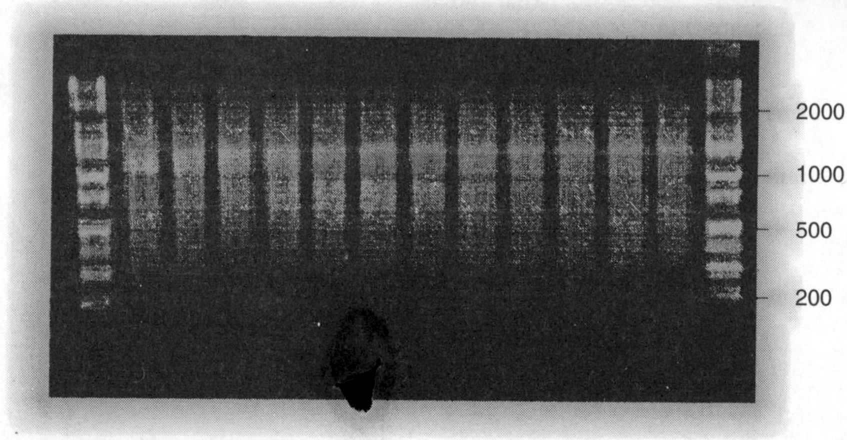

Figure 7.2: Sample image of the PCR amplification product run on a 2% TBE agarose gel. To check the PCR amplification step, 3 µL of each PCR product should be run on a 2% TBE agarose gel, at 120 V for 1 h. The PCR products should appear as smear with average size between 250 and 2000 bp.

that might remain on the bottom of the plate.

10. Add 50 µL EB buffer to each well. Cover the plate with PCR plate cover film.
11. oderately shake the MinElute plate on a plate shaker, e.g., Jitterbug (Boekel Scientific, cat. no. 130000), for 15 min.
12. Recover the purified PCR product by pipetting the eluate out of each well. For easier recovery of the eluates, the plate can be held at a slight angle. Make sure you take 50 µL off each well.
13. Take a 4-µL aliquot to measure the *concentration* of the purified PCR product using spectrophotometer.
14. Adjust the DNA concentration to 40 µg of per 45 µL solution using EB buffer (10 mM Tris-Cl, pH 8.5).
15. Transfer 45 µL of each purified DNA (40 µg) to new PCR tubes for the *fragmentation* reaction.

Process of Fragmentation

1. Take out the *fragmentation buffer, labeling reagent,* and TdT buffer and thaw on ice.
2. Preheat the PCR thermal cycler to 37°C.
3. Aliquot 45 µL of each sample into fresh PCR tubes. Put on ice.
4. Add 5 µL 10X Fragmentation Buffer to each sample in the tubes on ice and vortex at medium speed for 2 s (with tubes covered).
5. Get fragmentation reagent from the freezer and keep in enzyme block.
6. Prepare a dilution of the fragmentation reagent as listed in Table elsewhere in this chapter, based on the enzyme *concentration* on the stock tube.

Table 7.5: Fragmentation Reagent.

Fragmentation reagent stock concentration	2 U/µL	3 U/µL
Fragmentation reagent	3 µL	2 µL
10X Fragmentation buffer	15 µL	15 µL
Water	132 µL	133 µL
Total	150µL	150µL

7. Add 5 µL of diluted *Fragmentation Reagent* (0.04 U/µL) to each sample tube on ice. Pipet up and down several times to mix. *Perform* this step quickly.
8. Cover the tubes with strip caps and *seal tightly*.
9. Vortex the tubes at medium speed for 2 s, and spin briefly.
10. Place the fragmentation plate in preheated thermal cycler (37°C) as quickly as possible.
11. Run the following program: 37°C 35 min, 95°C 15 min, 4°C hold.
12. After the run, spin the tubes briefly.
13. Check the fragmentation product by run 4 RL of each fragmentation product on 4 % TBE agarose gel, at 120 V for approx 1 h. The product size should < 180 by. The recommended gel: 4 % TBE Gel, BMA Reliant precast.

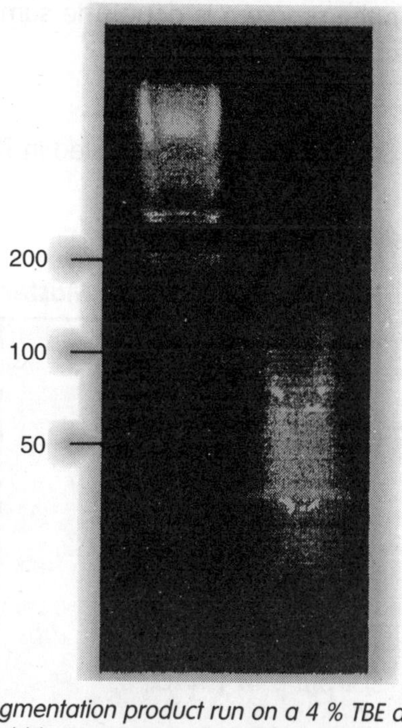

Figure 7.3: Sample image of the fragmentation product run on a 4 % TBE agarose gel. To check the fragmentation product, 4 µL of each reaction should be run on 4 % TBE agarose gel, at 120 V for approx 1 h. The product size should < 180 bp.

Table 7.6: Labeling Mixa[a].

5X TdT buffer	14 µL
GeneChip DNA labeling reagent	2 µL
TdT (30 U/µL)	3.5 µL
Total	19.5 µL

[a]Amounts for one reaction.

Labeling

1. Prepare Labeling Mix as *master mix* on ice as listed in Table elsewhere in this chapter and *vortex* at medium speed for 2 s. Prepare the mix for the number of samples you have plus 1.
2. Aliquot 19.5 µL of Labeling Master Mix into each tube containing 50.5 µL of fragmented DNA.
3. Cover the tubes with caps.
4. Vortex and spin briefly.
5. Put on the *thermal cycler* and run the following program (37°C 2 h, 95°C 15 min, 4°C hold).
6. Briefly spin the tubes when the program is done. The samples can be stored at -20°C if not proceeding to hybridization immediately.

Preparation of Hybridization

1. Prepare the Hybridization Cocktail Master Mix as listed in Table elsewhere in this chapter. Make 10 extra.
2. Aliquot 190 µL into Eppendorf tubes.

Table 7.7: Hybridization Cocktail Master Mix[a].

12X MES	12µL
100 % DMSO	13µL
50X Denhardt's solution	13 µL
0.5 *M* EDTA	3µL
10 mg/mL hsDNA	3µL
Oligonucleotide control	2 µL
Human Cot-1 (1 mg/mL)	3 µL
Tween-20 (3 %) 1 p L 5 *M* TMACL	140µL
Total	190 µl -

[a]Amount for one sample.

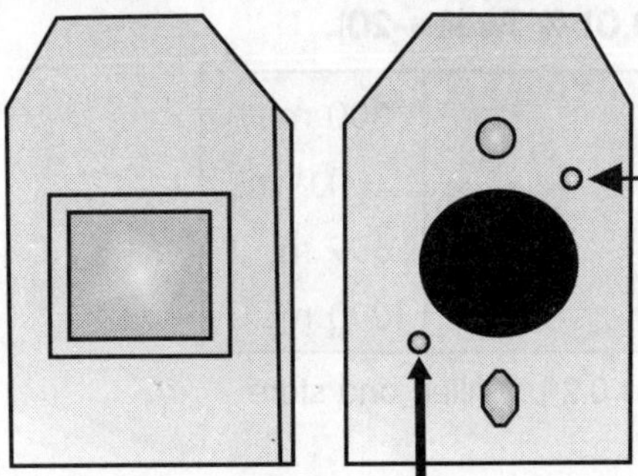

Figure 7.4: Illustration of loading the sample into GeneChip array cartridge.

3. Add 70 μL of labeled DNA into each tube. These samples can be used *immediately* or stored at -20°C.
4. Heat the 260 μL of hybridization mix and labeled DNA at 95°C in a heat block for 10 min to denature.
5. While it is heating, label the appropriate chips. Make sure to use the correct type of array(s). In this chapter, we will use GeneChip Mapping 50K Array Xba 240.
6. Cool down on ice for 10 s, and spin briefly in a microfuge.
7. Place the tubes at 48°C for 2 min.
8. Insert a fresh, P10 pipet tip into the upper septum on the array to vent it. Inject 200μL denatured hybridization into the array using an appropriate pipet. See Figure elsewhere in this chapter for illustration.
9. Hybridize overnight at 48°C for 16-18h at 60 rpm in the GeneChip Hybridization oven.
10. After hybridization, remove the hybridization cocktail from the *probe array* and set it aside in a microcentrifuge vial. Store at -80°C for long-term storage.
11. Fill the probe array completely with 250 μL of Array Holding Buffer. If necessary, the probe array can be stored in the Array Holding Buffer at 4°C for up to 3 h before proceeding with *washing* and *staining*. Equilibrate the probe array to room temperature before washing and staining.

Method of Staining

This section contains protocol of using the GeneChip Fluidics Station 450 to automate the washing and staining of GeneChip Mapping 50K Arrays. The *Fluidics Station* is controlled by GeneChip Operating Software (GCOS).

Please refer to the Fluidics Station User's Guide and the GCOS Instruction for the detailed instructions of installation, operation, *maintenance* of the Fluidics Station. Before start, register the experiment in the GCOS as New Experiment following the instruction of GCOS. Enter the information into the appropriate boxes. Then save the experiment in the *database*.

Table 7.8: Wash A: Nonstringent Wash Buffer (6X SSPE, 0.01 % Tween-20).

20X SSPE	300 mL
10 % Tween-20	1.0 mL
Water	699 mL
Total	1000 mL

Filter the buffer through a 0.2-μm filter, and store at room temperature.

1. Prepare the Washing Buffer A and Washing Buffer B as listed in Tables elsewhere in this chapter, respectively.
2. Prepare the following stock solutions that needed for making the buffers for staining.
   a. 0.5 mg/mL *Anti-Streptavdin* antibody: resuspend 0.5 mg *anti-Streptavdin* antibody in 1 mL of water. Store at 4°C.
   b. 12X MES Stock Buffer: make the 12X MES Stock Buffer (1.22 M MES, 0.89 M [Na+]) as listed in Table elsewhere in this chapter. Filter the buffer through a 0.2-μm filter. Do not autoclave. Store at 2 – 8°C, and shield from light. Discard the buffer if it turns to yellow.
3. Prepare the Stain Buffer Streptavidin phycoerythrin Solution Mix, Antibody Solution Mix, and 1X Array Holding Buffer for array staining.

Table 7.9: Wash B: Stringent Wash Buffer (0.6X SSPE, 0.01 % Tween-20).

20X SSPE	30 mL
10% Tween-20	1.0 mL
Water	969 mL
Total	1000 mL

Filter the buffer through a 0.2-μm filter, and store at room temperature.

Table 7.10: 12X MES Stock Buffer.

MES hydrate	70.4 g
MES sodium salt	193.3 g
Water	800 mL

Mix and adjust volume to 1000 mL. The pH should be between 6.5 and 6.7.

4. Priming the *fluidics* station: change the intake buffer reservoir A to Non-Stringent Wash Buffer, and intake buffer *reservoir* B to Stringent Wash Buffer. Run the Prime program.

Follow the instructions on the LCD.

5. In the Fluidics Station dialog box on the *workstation*, select the correct experiment name from the drop-down Experiment list. The Probe Array Type appears automatically.
6. In the Protocol drop-down list, select Mapping l00Kv1 450, to control the washing and staining of the probe array.
7. Choose Run in the Fluidics Station dialog box to begin the washing and staining. Follow the instructions in the LCD window on the *fluidics station.*
8. Insert the appropriate probe array into the designated module of the fluidics station while the cartridge lever is in the down or eject position. When finished, verify that the *cartridge* lever is returned to the up or engaged position.
9. When prompted to "Load Vials 1-2-3," place the three vials into the sample holders 1, 2, and 3 on the fluidics station.
   a. Place one vial containing 500 μL *streptavidin phycoerythrin* stain solution mix in sample holder 1.
   b. Place one vial containing 500 μL anti-streptavidin biotinylated antibody stain solution in sample holder 2.
   c. Place one vial containing 800 μL Array Holding Buffer in sample holder 3.

Table 7.11: Stain Buffer.

Water	666.7 μL
SSPE (20X)	300μL
Tween-20 (3%)	3.3 μL
Denhardt's (50X)	20 μL
Total (volume needed for one array)	990 μL

Table 7.12: Streptavidin Phycoerythrin Solution Mix.

Water	666.7 μL
Stain buffer	495 μL
1 mg/mL Streptavidin phycoerythrin	5.0 μL
Total	500 μL

   d. Press down on the needle lever to snap needles into position and to start the run.

Once these steps are complete, the fluidics protocols begin. The *Fluidics Station* dialog box at the *workstation* terminal and the LCD window *displays* the status of the washing and staining steps.

10. When staining is finished, remove the *microcentrifuge* vials *containing* stain and replace with three empty *microcentrifuge* vials as prompted.
11. Remove the probe arrays from the fluidics station modules by first pressing down the cartridge lever to the eject position.

12. Check the probe array window for large bubbles or air pockets.
    a. If bubbles are present, the probe array should be filled with Array: Holding Buffer manually, using a pipet. Take out one-half of the solution and then *manually* fill the probe array with Array Holding Buffer.
    b. If the probe array has no large *bubbles*, it is ready to scan on the GeneChip Scanner 3000. Pull up on the cartridge lever to *engage* wash block and proceed to array scan.
13. Shut down the Fluidics Station by running the ShutDown protocol.

Use of Scanner

This section contains protocol of using the GeneChip Scanner 3000 to scan the probe arrays. The scanner is also controlled by GCOS. Please consult *appropriate* manuals for the detailed instructions of *installation, operation, maintenance* of the GeneChip Scanner.

Table 7.13: Antibody Solution Mix.

Stain buffer	495 µL
0.5 mg/mL Biotinylated antibody	5.0 µL
Total	500 µL

Table 7.14: 1 X Array Holding Buffer (100 mM MES, 1 M [Na+], 0.01 % Tween-20)[a]

12X MES Stock buffer	8.3 mL
5 M NaCl	18.5 mL
10 % Tween-20	0.1 mL
Water	73.1 mL
Total	100 mL

[a]Store at 2 - 8°C, and shield from light.

Make sure the laser is warmed up prior to scanning by turning the GeneChip Scanner 3000 laser on at least 10 min before use. If the array was stored at 4°C, allow to warm to room *temperature* before scanning. After completing the procedures described in this chapter, the scanned probe *array image* (.dat file) is ready for analysis.

1. Go to the scanner window in the GCOS. Select the experiment that *registered* at the beginning of the array washing.
2. Click Load/Eject button on GCOS to load the array to be scanned. Load the array by open the sample door of the scanner and insert the array into the holder.
3. Click Start. Then click the OK in the Start Scanner dialog box to start the scanning.

The scanned array image will appear on the screen as scan progresses.

Analyzing the Results

The raw data (scanned array image, which is saved as dat file) acquired using the GCOS will be analyzed using GTYPE to generate *genotyping* calls for each SNP markers on the array. The outline of the data analysis workflow is shown in Figure elsewhere in this chapter.

Table 7.15: FS-450 Fluidics Protocol - Mapping100Kv1_450.

Post Hyb Wash no. 1	6 cycles of 5 mixes/cycle with Wash Buffer A at 25°C
Post Hyb Wash no. 2	6 cycles of 5 mixes/cycle with Wash Buffer B at 45°C
Stain	Stain the probe array for 10 min in SAPE solution at 25°C
Poststain wash	6 cycles of 5 mixes/cycle with Wash Buffer A at 25°C
Second stain	Stain the probe array for 10 min in antibody solution at 25°C
Third stain	Stain the probe array for 10 min in SAPE solution at 25°C
Final wash	10 cycles of 6 mixes/cycle with Wash Buffer A at 30°C. The final holding temperature is 25°C
Filling array	Fill the array with Array Holding Buffer

In this section, we will provide a brief introduction of the data analysis for GeneChip Mapping array using GTYPE. For advanced data analysis, please refer to Affymetrix GeneChip Genotyping Analysis Software User's Guide (www.affymetrix.com).

1. Open the Batch Analysis window. Then, select the appropriate array data (cel files) for analysis by *dragging* the files into the Batch Analysis window.
2. Start the analysis by clicking Analyze and the selected files will be analyzed.
3. After Batch Analysis is complete a report will be displayed *summarizing* data from the samples. The report contains information that can be used to assess the *performance* of the sample, such as call rate. In general, a successful experiment should have a call rate of >90 %.
4. Following analysis, the chp files are generated for each sample. The chp files can be

GCOS

1) Experiment registration
2) Washing and staining arrays
3) Scanning arrays
4) Automatically generate. CEL files

GTYPE

5) Batch analyze data
6) Generate report
7) Generate. CHP files
8) Exprot SNP genotyping data from. CHP files for additional analyses

Figure 7.5: Data analysis flow for GeneChip Mapping 50K Arrays when using GeneChip Operating Software and GeneChip Genotyping Analysis Software.

opened by *double-clicking* on the files in the GTYPE data file tree. Multiple .chp files can be opened *simultaneously.*

Once opened the chp files will display a *Dynamic Model scatter plot* in the upper portion of the window and a data table in the lower portion of the window as shown in Figure elsewhere in this chapter. In the data table a *genotype call* for each SNP along with a confidence score is displayed.

5. The SNP call data can be exported for use in third party software products in either tab delimited file format, *Merlin format,* GeneHunter format, or Microsoft Excel format. To export data:

   (a) Open the chp files you would like to export in GTYPE.

   (b) Select Export and Table in the menu bar.

   (c) Specify the location and format of the export data.

POINTS TO REMEMBER

1. One major reason of reduced genotyping accuracy is contamination. The most likely potential source of *contamination* for SNP *mapping array* assay is the previously amplified PCR product. It is recommended to set up the laboratories as pre-PCR Clean Room, PCR Staging Room, and Main Lab. The pre-PCR Clean Room and PCR Staging Room can be *substitute* with a PCR clean hood and a PCR *Staging hood,* such as the AirClean 600 PCR Workstation (AirClean Systems).

   To avoid the potential *cross-contamination,* each step of the SNP array mapping assay should performed in the dedicated room/area as illustrated in Table elsewhere in this chapter. It is also highly recommended to have dedicated *equipments* for each dedicated roomlarea, such as *pipet, vortexer, desktop centrifuges, thermal cycler,* and so on.

2. The quality of the input genomic DNA is critical for the success of the GeneChip Mapping Assay. DNA must be double-stranded (not single-stranded) and free of PCR inhibitors (such as high concentrations of heme and EDTA). DNA must not be *contaminated* with other human genomic DNA sources, or with genomic DNA from other organisms. DNA must not be highly degraded. The *recommended* genomic DNA isolation methods are:

   (a) SDS/ProK digestion, *phenol-chloroform* extraction, Microcon or Centricon (Millipore) *ultrapurification* and *concentration.*

   (b) QIAGEN; QIAamp DNA Blood Maxi Kit.

3. The most common problem encountered for this SNP array-based *mapping* assay is the low SNP call rate. The followings are potential causes and respective recommendations.

   (a) Over fragmentation of DNA sample due to incorrect dilution of Fragmentation *Reagent* (DNase I) stock. Use correct concentration of Fragmentation Reagent (DNase I). Check concentration (U/μL) on the label of the *Fragmentation* Reagent stock and prepare the *fragmentation dilution* as described elsewhere in this chapter. Work quickly and on ice, transfer reaction tubes to preheated *thermal cycler* (37°C).

Table 7.16: Laboratory Setup.

	Assay steps	Template (genomic DNA)	PCR product
Pre-PCR clean room	Step 1: Reagent preparation	–	–
PCR staging room	Step 2: Enzyme digestion		
	Step 3: Ligation	+	–
	Step 4: PCR (set up)		
Main lab	Step 4: PCR (reaction)		
	Step 5: PCR clean up		
	Step 6: Fragmentation	+	+
	Step 7: Labeling		
Microarray core	Step 8: Hybridization		
	Step 9: Washing, staining, and scanning	+	+

(b) Failed labeling reaction. Use a new vial of *Terminal Dideoxy Transferase*. Verify the labeling reagents and repeat labeling.

(c) Suboptimal quality of the input genomic DNA. Ensure DNA samples are of high quality. Genomic DNA that has been preamplified using certain methods (e.g., Phi29 DNA polymerase-based genome amplification) may give reasonable results. Pilot test should be planned to evaluate *preamplification* of specific DNA samples with this assay.

(d) Mix up the samples. Repeat the assay using correct samples.

Chapter 8

Identifying Proteins

Developments in *mass spectrometry* technology, together with the availability of extensive DNA and protein sequence databases and software tools for *data mining*, has made possible rapid and sensitive mass spectrometry-based procedures for protein identification. Two basic types of mass spectrometers are commonly used for this purpose; Matrix-assisted laser *desorption/ ionization* (MALDI)-time-of-flight (TOF) mass spectrometry (MS) and *electrospray* ionization (ESI)-MS. MALDI-TOF instruments are now quite common in *biochemistry* laboratories and are very simple to use, requiring no *special training*.

ESI instruments, usually coupled to *capillary*/nanoLC systems, are more complex and require expert *operators*. We will therefore focus on the use of MALDI-TOF MS, although the sample *preparation* is identical for both methods. The *principle* behind the use of MALDI-TOF MS for protein identification is that the digestion of a protein with a specific *protease* will generate a mixture of peptides unique to that protein.

Measuring the molecular masses of these peptides then gives a characteristic dataset called a peptide mass fingerprint (PMF). The PMF data can then be compared with *theoretical peptide* molecular masses that would be generated by using the same *protease* to digest each protein in the sequence database, to find the best match.

Provided the protein being analyzed is present in the database being searched and the data are of *sufficient quality*, the best match should be the correct protein. In order to judge the validity of a protein identification by this method, some means of scoring the quality of the match must be used. The procedure described here involves cutting protein bands or spots from one-dimensional (1D) or two-dimensional (2D) *polyacrylamide* gel *electrophoresis* (PAGE) gels, *de-staining* the gel pieces, reducing and *alkylating* the protein, digesting with *trypsin*, using MALDI-TOF MS to determine the masses of the *tryptic peptides*, and database searching with the PMF data to identify the protein.

The sample-processing steps can be performed in microfuge tubes or in 96-well plates. One person can easily process a 96-well plate in a day, but if higher *throughput* is required, each step can be automated, allowing the *possibility* of several hundred protein *identifications* per day. *Spot cutting, sample processing,* and sample plate loading robots are *commercially* available and routinely used in many high-throughput laboratories.

REQUIREMENTS

1. 0.5-mL microfuge tubes (or 96-well V-bottom *polypropylene microtiter* plates).
2. One Touch Spot Picker (1.5 mm) (The Gel Company) optional.
3. *Silver de-staining* solution (for silver-stained gels only)-dissolve *potassium ferricyanide* (2 mg/mL) in *sodium thiosulphate* solution (0.2 mg/mL). Make fresh immediately before use.
4. Sonicator bath (or vortex mixer with attachments for unattended use with 0.5-mL microfuge tubes or 96-well microtiter plates).
5. Aqueous buffer: 50 mMNH_4HCO_3 make fresh weekly.
6. Organic buffer: 50 mMNH_4HCO_3/acetonitrile 1: 1 make fresh weekly.
7. DL-*dithiothreitol* ultra pure (DTT) solution: 10 mM DTT in aqueous buffer—make fresh immediately before use.
8. *Iodoacetamide* solution (ultra pure): 50 mM *iodoacetamide* in aqueous buffer-make fresh immediately before use. *Note: iodoacetamide is toxic!*
9. *Trypsin solution* (Modified Sequencing Grade, Promega).
10. 10% *trifluoroacetic acid* (TFA) in water make fresh weekly.
11. Centrifugal vacuum concentrator.
12. Matrix solution.
13. Peptide *calibration* mixture.
14. MALDI sample plate.
15. MALDI-TOF mass *spectrometer.*

PROCEDURES

Protein identification by PMF is usually performed on protein bands or spots cut from 1 D or 2D PAGE gels, so this is the method described here; however, it can also be applied to proteins in solution with minor *modifications.* One of the biggest problems that may be *encountered,* particularly when low amounts of protein are analyzed, is *contamination.* describes elsewhere in this chapter some *precautions* to minimize the effects of contamination.

All washing steps during the processing of the gel pieces can be done either on a vortex mixer or in a sonicator bath.

De-Staining Technique

Ideally, the gel should have been stained with a Coomassie® stain-preferably a colloidal Coomassie if detection sensitivity is an issue, e.g, or one of the commercially available stains. This method is also compatible with Sypro® stains (though an ultraviolet transilluminator will be required to visualize the protein bands or spots during *excision*). Standard silver stains are not compatible, but certain modified silver stains e.g, or one of the *commercially* available *"mass spectrometry compatible"* silver stainscan be used; however, even these usually give inferior mass spectrometry data compared to Coomassie- or Sypro-stained gels.

Removal of Gel Piece

1. Place the stained gel in a disposable Petri dish on a light box and cut out the spot/ band with a manual spot picker or a clean scalpel/razor blade, without taking any excess gel.
2. If necessary, cut the gel spot/band into 1-mm pieces and transfer into a 0.5-mL microfuge tube or one well of a 96-well microtiter plate.

De-Staining Coomassie- or Sypro-Stained Gel Pieces

1. Wash with aqueous buffer (100 μL) 1 ↔ 5 min.
2. Wash with organic buffer (100 μL) 2 ↔ 5 min.

De-Staining Silver-Stained Gel Pieces

1. Wash with silver de-staining solution (100 μL) until de-stained (approx *5-30 min)* the gel will retain a pale yellow colour.
2. Wash with high-purity water (100 μL) 2 ↔ 5 min.
3. Wash with aqueous buffer (10μL) 1 ↔ 5 min.
4. Wash with organic buffer (100 μL) 1 ↔ 5 min.

Reduction of Disulfide Bonds

Reduction of disulfide bonds followed by alkylation of the free cysteines to prevent reoxidation, while not essential for the digestion of most proteins, generally gives better results. This is due to increased susceptibility of the *reduced/alkylated* protein to tryptic digestion and the absence of any *disulphide*-linked peptides (which are not matched in the database search) from the PMF data.

1. Incubate de-stained gel pieces in DTT solution (100μL) 1 h at 50% °C in an oven. If using a 96-well plate, use sealing film or a sealing lid, to exclude oxygen and prevent drying out.
2. Cool to room temperature, remove, and discard DTT solution. Add iodoacetamide solution (100 μL) and incubate with occasional mixing (vortex) for 1 h in the dark at room temperature.
3. Discard supernatant and wash gel pieces with aqueous buffer (100 μL) for 5 min, then with organic buffer (100 μL) for 2 H 5 min.
4. Dry the gel pieces completely in a centrifugal vacuum concentrator. Caution should be exercised in handling the dried gel pieces, as they are easily lost from tubes or plates.

Enzymatic Hydrolysis of Protein

1. Take an aliquot of trypsin (10 μL of 10X stock solution) from the freezer, add aqueous buffer (90 μL) and mix.
2. Rehydrate gel pieces for 10 min in trypsin solution (approx 1 μL per mm^3 gel). There should

be little or no excess liquid after rehydration.

3. Add an equal volume of aqueous buffer and incubate at 37°C overnight (or for at least 3 h). Use an incubator, not a heating block. If using a 96-well plate, use sealing film or a sealing lid, to prevent drying out.
4. Add one-tenth volume of 10% TFA and sonicate/vortex for 5 min.
5. The resulting supernatant is used directly for MALDI-TOF MS.
6. Residual peptides can be washed from the gel piece if required with a small amount of 0.1 % TFA.

Observing Peptides by MALDI-TOF MS

The quality of the MALDI-TOF spectrum that will be obtained from the sample depends crucially on the sample/matrix preparation. The basic requirements are for a uniform *microcrystalline* layer of matrix/sample and the removal of salts and other *contaminants*. There are numerous *published methods*, but the one described here is robust and quite simple, requiring no precleanup steps.

For the best results, a high-quality matrix, as supplied by MALDI-MS manufacturers or in a commercial PMF kit, should be used. Alternatively, analytical-grade matrix can be recrystallized. MALDI-TOF spectra must be *calibrated* in order to achieve sufficient accuracy for database searching. This is done by acquiring spectra on peptide standards to generate a *calibration* curve, which is applied to the *experimental data.*

The calibration peptides can be analyzed separately from the experimental sample (*external calibration*), or they can be mixed with the experimental sample (internal calibration). Internal calibration is more accurate (typically 10-20 ppm for a strong spectrum) than external (typically 100 ppm or more), but is more difficult, as the amount of standard peptides used needs to be matched to the level of the experimental sample peptides.

The addition of too much peptide standards can suppress the signal of the sample peptides and vice versa. A commonly used variation of the internal *calibration* method utilizes the *autodigestion* fragments of trypsin (m/z 842.5094 and 2211.1040 for porcine trypsin) for calibration, instead of adding additional peptides.

Sampling

1. Apply matrix solution (0.5 µL) to a clean MALDI plate and allow to dry (sample spot).
2. Apply protein digest (0.2-2 µL) to the sample spot and allow to dry. If a large amount of gel was used for the digestion, the *supernatant* volume could be much larger than 2 µL (up to 20 µL). It is not usually necessary to use all of the supernatant unless the protein band was quite weak. In this case, the entire supernatant can be concentrated to 1-2 µL and used.
3. Apply 0.1 % TFA (5-10 µL) to each sample spot, leave for 30 s, *remove*, and *discard*, then *repeat*; this step desalts the sample.
4. If using external calibration, apply 0.2 µL of peptide calibration mixture as close as possible to (but not touching) each sample spot and allow to dry (*calibrant spot*).

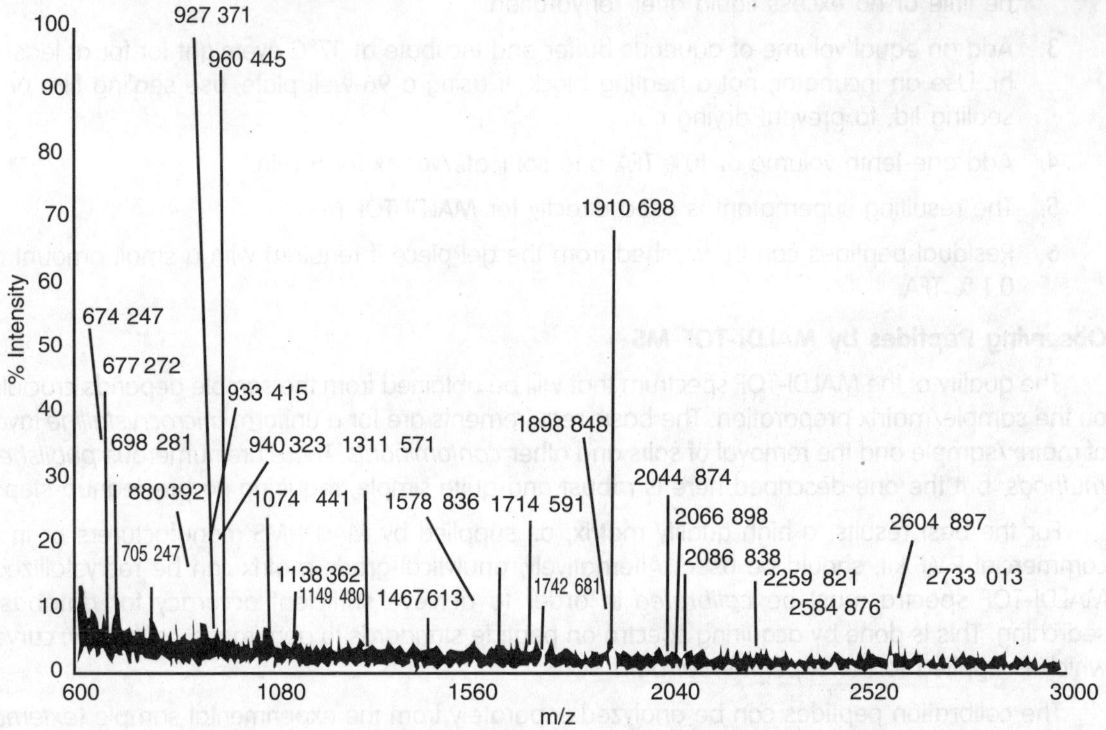

Figure 8.1: Matrix-assisted laser desorption/ionization (MALDI)-time-ot-flight (TOF) spectrum of a trypsin-digested one-dimensional gel band. Peaks are labeled with their monoisotopic masses. Note that these are not the masses of the peptides, but of the peptide (pseudo)molecular ions. In MALDI spectra, peptide molecular ions arise predominantly through the addition of a proton to the peptide, giving a mass increase of 1.007 Da. The molecular ions are usually denoted as MH^+ or [M+H]'.

Estimation of Peptides Masses

The precise operation of the MS is instrument dependent, but the basics are very similar. Increasing the laser power increases the signal, but also decreases the *resolution* and *"burns*" off the sample faster. Therefore, the laser power should be set to the lowest level that gives a good signal. Set the mass range (e.g., m/z 600-3500) and the number of shots to be acquired (100-200 is reasonable), and set the laser power low.

Gradually increase the power until an even distribution of noise appears across the whole mass range. Within a few shots, peaks should begin to appear above the noise. Continue to adjust the laser power until the signal level is satisfactory. The *heterogeneous* nature of the *matrix surface* means that some areas of the sample spot will give better *peptide signals* than others.

The difference can be dramatic, so it is a good idea to *periodically* move the laser position, while assessing the signal intensity, to find the hot spots. If external calibration of the spectrum is to be used, a spectrum of the calibration standards should be acquired immediately, using the same power setting. Figure shows elsewhere in this chapter a typical MALDI *spectrum* obtained from the trypsin digestion of a weak colloidal *Coomassie-stained* I D gel band.

Finding Protein Databases

The PMF data must now be extracted from the MALDI spectrum using the appropriate software associated with the MS instrument used. The baseline threshold should first be adjusted to ensure that all of the peptide signals are detected, without including any noise. If the spectrum has been obtained on a high resolution (reflectron) instrument, the peptide signals will appear as multiple peaks separated by 1 Da, owing to the presence of ^{13}C isotopes in the peptides.

"*De-isotoping*" and "*centroiding*" of the MS data is necessary, so that only the monoisotopic peptide masses are included in the PMF data. The MS data analysis software will have an option for the data to be shown as a list of masses (peak list), which can be saved as a text file or simply copied directly into the data entry field off the *database* search engine.

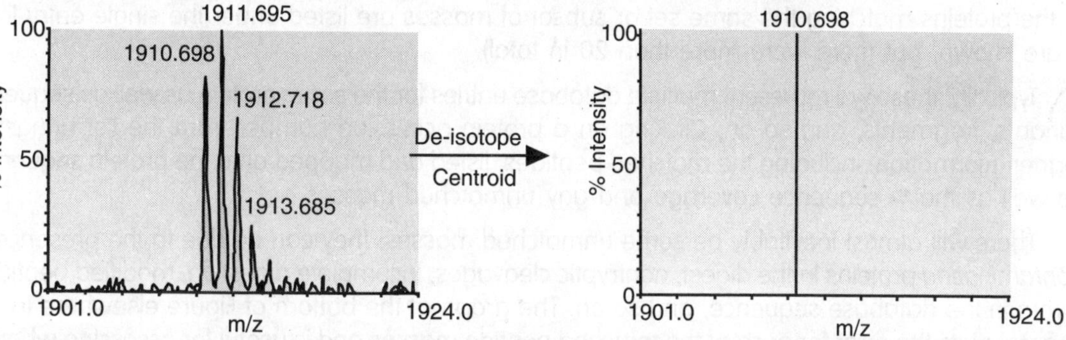

Figure 8.2: Expanded view of the peptide signal at around m/z 1910 from the Matrix-assisted laser desoiption/ ionization (MALDI)-time-of-flight (TOF) spectrum in Figure 8.1. De-isoping removes the ^{13}C isotope peaks and centroiding reduces the remaining peak to a single data point.

There are a number of search engines that can be used for PMF searches, some of which are freely available on the Internet. We will use Mascot to demonstrate a search, but all have essentially the same functions.

1. Select the database to be searched from the drop-down list, e.g., NCB1nr.
2. If taxonomic information is known for the sample origin, select from the list; otherwise, select "*All entries.*"
3. Select "trypsin" as the enzyme.
4. "Allow up to" 1 missed cleavage.
5. Select "*Carbamidomethyl* (C)" as a fixed modification.
6. Select "Oxidation (M)" as a variable modification.
7. If the MALDI spectrum was internally calibrated, the peptide tolerance can be set to 0.1 Da; otherwise. 0.3 to 0.5 Da is a reasonable starting point.
8. Select "MH "for Mass values.
9. Select "*Monoisotopic*" if the MALDI data were acquired in reflectron mode, or "*Average*" for data acquired in linear mode.
10. If the peak list was saved as a text file, browse to the file location; otherwise, paste the

list directly into the "Query" field.

11. Select "Auto" from "Report top" hits.
12. Start search.

Within a few seconds, the search result will appear, which should look similar to Figure elsewhere in this chapter. This shows a graphic representation of the results and a list of significant matches. A single significant match is indicated by the graph with a maximum score of 240, well above the significance threshold of 63.

The score is related to the probability that the match is real rather than purely random, and is also expressed as an expect value equivalent to the BLAST E-value, 1.2e-19 in this case, indicating that this is almost certainly a genuine match. In the "Concise Protein Summary Report" shown, all of the proteins matching the same set or subset of masses are listed under the single entry (only 5 are shown, but there were more than 20 in total).

Typically, these will represent multiple database entries for the same protein as well as sequence variants, fragments, and so on. Clicking on a protein accession number from the list brings up further information, including the matched peptides, listed and mapped onto the protein sequence, as well as the % sequence coverage and any unmatched masses.

There will almost inevitably be some unmatched masses; they can be due to the presence of *contaminating* proteins in the digest, nontryptic cleavages, incomplete digestion, modified peptides, errors in the database sequence, and so on. The graph at the bottom of Figure elsewhere in this chapter plots the error for each of the matched peptide masses and is useful for assessing whether the *tolerance* setting used for the search was appropriate. If *desired*, a more appropriate setting could be chosen and the data researched.

Weak Points

Some of the most common reasons why a database search may not give a significant match. are:

1. Too few masses in the PMF data: depending on the size of the database and the parameters settings, a minimum of four to six matched peptides are required for a significant hit. This situation could arise simply because the *spectrum* is very weak, and so only a few of the *strongest* peptide signals are visible, or because the particular protein does not yield many peptides within the useful mass range (e.g., small proteins, proteins with very high or very low numbers of *potential trypsin* cleavage sites, proteins that are resistant to trypsin digestion).
2. More than one protein present in the sample: with good-quality PMF data, mixtures containing two to three proteins can be identified, but the presence of peptide masses from the other proteins reduces the score for each individual match. With *lower-quality* data, this can result in all of the scores dropping below the significance threshold. The presence of contaminants, e.g., *trypsin* or *keratins*, has the same effect; however, known peptide masses derived from *keratins* and *trypsin* can be removed from the PMF data prior to *searching*.
3. Peptide tolerance set too low or too high.

4. Incorrectly calibrated spectrum.
5. Searching EST databases: ESTs are generally short and therefore do not usually provide *sufficient numbers* of matched peptides for a significant score.
6. The protein is not present in the database being searched.

POINTS TO REMEMBER

1. Contamination is a serious concern when attempting to identify low amounts of protein by MALDI MS. Use only the highest quality reagents, high-performance liquid chromatography (HPLC)-grade solvents, and high-quality plasticware to minimize non-protein contamination. Ideally, a set of glass/plasticware and reagents/ buffers should be dedicated for MALDI-MS analysis. Laboratory dust is a major source of protein contaminants (keratins), and so every effort must be made to exclude dust from samples, buffers, and reagents.

   Communal buffers, stains, de-stains, and so on, are often contaminated with dust and should be avoided. Rinse all glassware with high-quality water before making up buffers. Wear nitrile gloves (latex contains protein contaminants) and a lab coat. Keep a cover over the gel during staining.

   Never stain a gel in a container that has been used for processing Western blots, as they will be contaminated with blocking proteins. Handle the gel only if absolutely necessary, and then avoid touching parts of the gel that are to be cut.

   Gel pieces can be cut and processed in laminar-flow cabinets, although this is usually not necessary if sensible precautions are taken.
2. Modified trypsin is preferred for protein digestion, as it is less susceptible to autodigestion. The trypsin lox stock solution is prepared by dissolving the trypsin (20 µg) in the solvent supplied with the enzyme (50 mMacetic acid; 200 µL). This is stored at -70°C in aliquots (10 µL) and is stable for at least l yr.
3. MALDI analysis kits (e.g., Sequazyme peptide mass standards kit from Applied Biosystems) are a convenient way to obtain matrix, peptide standards, and solvents; however, these items can easily be purchased *independently*. Any peptides that span a reasonable proportion of the useful mass range can be used for calibration, provided their accurate molecular weights are known.

   α-Cyano-4-*hydroxycinnamic acid* is the most common matrix used for peptide analysis. Matrix solution is prepared by dissolving 5 mg of matrix in 1 mL of 4:1 acetonitrile/water containing 0.1 % TFA.

   A mixture of peptide calibration standards-e.g., des-Arg'-bradykinin (monoisotopic mass 904.4681), angiotensin 1 (monoisotopic mass 1296.6853), ACTH (1-17 clip) (monoisotopic mass 2093.0867), ACTH (18-39 clip) (monoisotopic mass 2465.1989)-is made to a concentration of 2 pmol/µL each in 0. 1°/0 TFA in water. This solution is then mixed 1: 1 with matrix solution to give the final peptide-calibration mixture.
4. Larger volumes of digests can be concentrated by vacuum centrifugation, but this can

lead to significant peptide losses, particularly if the sample is concentrated to dryness. Alternatively, peptides can be concentrated and desalted by using a microscale pipet-tip format solid-phase extraction device (e.g., micro C 18 Zip Tips from Millipore, Omix C I 8MB tips from Varian, or STAGE tips from Proxeon). The details of their use in desalting peptide solutions for MALDI analysis are given with the product, but for the final elution step, use the matrix solution described above, to elute the peptides directly onto the MALDI plate.

5. Freely available PMF search engines include: Mascot from Matrix Science (http://www.matrixscience,com/cgi/search form.pl?FORMVER=2&SEARCH=PMF), MS-Fit from Protein Prospector (http://prospector.uesf.edu/ucsfhtml4.0/msfit. htm), and PeptIdent from ExPaSy (http://us.expasy.org/tools/peptident.html).

6. SwissProt is a relatively small but highly annotated database with minimal redundancy. NCBInr and MSDB are much larger, but have multiple entries for many proteins. Other search engines may have additional database choices, and in-house copies allow custom databases to be used.

   Specifying the taxonomy as closely as possible reduces the number of database entries that need to be considered in the search. This reduces search times, but more importantly, reduces the threshold score for significant matches, increasing the confidence of any protein identification.

   The cleavage specificity of trypsin is C-terminal to Lys and Arg residues (except where followed by Pro). However, not every such peptide bond will be cleaved. The number of missed cleavages to consider can be specified, but increasing the number decreases the significance of any match and should normally be set to one.

   Fixed modifications are modifications to specific amino acids that are considered to be complete-i.e., every occurrence of the amino acid in the sequence is assumed to carry the modification, and the unmodified amino acid is not considered. Variable modifications, on the other hand, are incomplete, and therefore both the modified and unmodified amino acid are considered in the search. In the example discussed above, the reduction/alkylation should result in complete carbamidomethylation of all cysteine residues; thus, "Carbamidomethyl (C)" was chosen as a fixed modification, whereas methionine oxidation, a common artifactual modification that is usually incomplete, was selected as a variable modification. The use of multiple variable modifications will greatly reduce the significance of any match and should therefore be used with caution.

   Most PMF search engines allow a mass for the protein to be entered; the search will then consider only database entries within a window around this mass. This can be dangerous, as proteins are subject to processing/degradation, which can significantly increase or decrease their molecular mass. Mascot uses a more complex method of applying this parameter, but in most cases this can be left blank.

   The peptide tolerance is a window around each mass value in the peak list within which a theoretical database peptide mass must fall, in order to be matched.

Chapter 9

Non-Coding RNA

It is now well-acknowledged that *noncoding* RNAs play an essential role in many cellular processes (e.g., *protein synthesis, regulation*), even if the function of the majority of RNAs remains to be elucidated Many of noncoding RNAs have characteristic secondary structures that are highly conserved in evolution. Identifying conserved structure is the first step toward the comprehension of the function of the molecule. *Computational* approaches provide *unexpansive* and *efficient* tools for that purpose.

From a *historical* perspective, there are two main *complementary* approaches to address RNA folding prediction: *thermodynamic* models and *phylogenetic* models. The secondary structure of an RNA molecule depends of the formation of *basepairings*: Watson-Crick (A-U and G-C), wobble (G-U), or even noncanonical pairings. In the *thermodynamic* approach, the fundamental assumption is that the molecule adopts a *globally* minimum free energy structure.

Negative *stabilizing* energies are assigned to the stacking of basepairs in helices and *destabilizing* energies are assigned to *unpaired* elements, such as bulge or *multibranched* loops. In this model, the folding problem amounts to searching for the set of base pairs that minimizes the free energy level. This *strategy* is implemented in the Mfold and RNAfold programs. The main limitation of these methods is that the right structure ma be overwhelmed by a large number of potential structures having *equivalent*, or even better, energy level.

Furthermore, there is presently no thermodynamic parameters that deal with *pseudoknots* within this model. The other line of research for structure inference is *phylogenetic* analysis. The main idea of this approach is to extract information from the *similarities* and differences between different, but *homologous*, RNA sequences. *Phylogenetic* analysis relies on the assumption that the spatial structure of a molecule is more highly *conserved* than is its sequence. In other words, the sequence is free to change during evolution.

In terms of secondary structure, this means that mutation of a base involved in a pairing should generally be compensated by a change in its pairing partner. This guarantees that the ability of the both bases to form *isosteric basepairs* is retained. This phenomenon is called covariation, or *compensatory* mutation. If *sufficient* numbers of sequences are available, these covariations can be identified *statistically* directly from a multiple sequence *alignment*. The list of structures determined by comparative analysis is long: ribosomal RNAs, transfer RNAs, RNase P RNAs, HACA box RNAs, snoRNAs, and so on. The drawback of pure *plylogenetic* approaches is that they need a large number of related sequences (more than 10) to be *theoretically* sound. Furthermore, the accuracy of the result strongly depends on the quality of the *multiple alignment*. *Automatically* aligning RNA sequences is a difficult issue.

The purpose of the caRNAc software is to achieve more flexibility than pure *comparative* methods by combining both thermodynamic and *phylogenetic* information. caRNAc does not require any prior alignment between sequences. This *implies* that it can successfully handle sequences with low level of *conservation* (from 60%).

AVAILABILITY

caRNAC is available on a website. All that is needed is a W3C compliant web *browses*.

Frequent users also may download the *platform* and *install* it locally. CaRNAc requires a C compiler. All source codes are available on the *website*.

HANDLING

Choose the "*web server*" section in the main menu. The "*examples*" section provides several data sets and commented results.

The input submission form contains three main fields.

Enter a name for the sequence (optional)

This name serves as a label for the output page.

```
>Bacteriophage T4 Thr–tRNA

GCUGAUUUAGCUCAGUAGGUAGAGCACCUCACUUGUAAUGAGGAUGUCGGCGGUUCGAUUCCGUCAAUCAG

CA

> Yeast   (S . cerevisiae)  mitochondrial  phe–tRNA

GCUUUUAUAGCUUAGUGGUAAAGCGAUAAAUUGAAGAUUUAUUUACAUGUAGUUCGAUUCUCAUUAAGGGC

A

> Halobacterium  volcanii   phe – t RNA.

GCCGCCUUAGCUCAUACUGGGAGAGCACUCGACUGAAGAUCGAGCUGUCCCCGGUUCGAAUCCGGGAGGGG

GCA
```

Figure 9.1: Comparative sequences of tRNAs.

Enter the RNA sequences

The data set should include at least two distinct RNA sequences, and sequences should be in FASTA format. A sequence in FASTA format consists of a *single-line description*, followed by lines of sequence data. The first character of the description line is a greater than (">") symbol in the first column. Figure elsewhere in this chapter gives an example with three tRNA sequences, that will serve as a guideline for the remaining of this *presentation*. All nonalphabetic characters are removed. IUPAC symbols are not *supported*. Sequences may be pasted or uploaded from a file.

Enter an e-mail address

This address is used to send the identifier of the job to the user once the job is completed.

The form then proposes several parameters that determine the final predictions. The default values lead to the most reliable results in average.

Eliminate redundant sequences

By default, caRNAc discards sequences that are too close (more than 98 % of identity). Uncheck the box to fold all the sequences.

Take GC content into consideration

When this option is selected, caRNAc uses variable energy thresholds for stems according to the average GC percent of the involved sequence.

Allow isolated stems

When this option is selected, caRNAc permits the creation of stem in one sequence alone, without any counterpart in any other sequence. This option may give better results if there is a large evolutionary distance between *structures*, or when the sequences are of radically different lengths. But it is time and space *consuming*. So it should be *selected* with *caution*.

The folding is launched by pressing the "RUN" button. Each job is assigned a unique identifier (ID). The *computation* of the putative foldings ranges from a few seconds for short sequences (less than 300 nt) to several minutes for longer sequences (up to 2000 nt).

When the job is completed, the results are displayed on a new page, and an alert e-mail is sent to the user. Results are available for 24 h and may be retrieved with the ID using the "retrieve a result with an ID" section in the main menu.

Output Page

For each sequence, the predicted secondary structure is given in five formats. Note that all structures need not to be identical: the program is robust to minor variations in the structure between the sequences.

Connect notation (ct)

It provides a textual description of the *basepairings*. The syntax is as follows: columns 1, 3, 4, and 6 redundantly give sequence indices, column 2 gives the sequences and column 4 gives "j" in position "i" if "(i,j)" is a *basepair*, otherwise this is zero. The heading of the file contains the size of the sequence and its name (found in the FASTA sequence).

A Connect notation (extract)

74 Halobacterium volcanii Phe–tRNA.

1	G	0	2	73	1
2	C	1	3	72	2
3	C	2	4	71	3
4	G	3	5	70	4
5	C	4	6	69	5
65	G	64	66	0	65
66	G	65	67	8	66
67	A	66	68	7	67
68	G	67	69	6	68
69	G	68	70	5	69
70	C	69	71	4	70
71	G	70	72	3	71
72	G	71	73	2	72
73	C	72	74	1	73
74	A	73	0	0	74

B JPEG file (or POSTSCRIPT file)

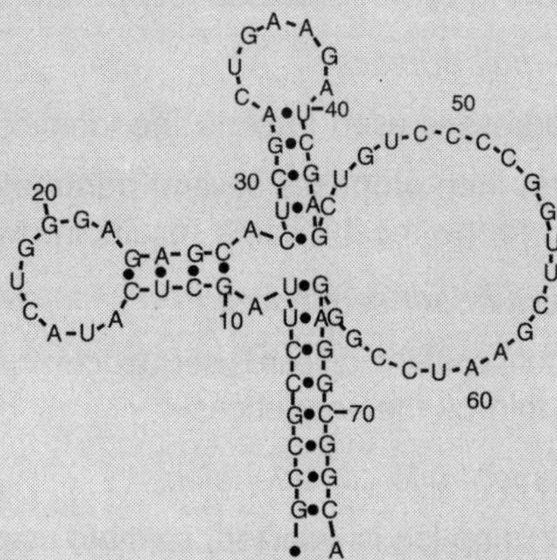

C List of contraints

F 1 73 8

F 10 26 4

F 28 44 5

D Bracket notation

GCCGCCUUAGCUCAUACUGGGAGAGCACUCGACUGAAGAUCGAGCUGUCCCCGGUUCGAAUCCGGGAGGCGGCA

((((((((.((((..........)))).(((((.......))))).......................)))))))).

Figure 9.2: Showing output formats for the third tRNA composed of three helices.

Jpeg file

This file is generated from the CT file using the freely distributed drawing tool Naview. It contains a graphical two-dimensional representation of the secondary structure.

Postscript file

This is a conversion of the Jpeg file to the postscript format. This format is ready-to-print.

List of constraints

This text file gives an *equivalent* formulation of the structure. Each line contains the specification of one stem: "F i j k" means that there is an helix of length k formed between the positions [i,i+k-1] and [j-k+1,j]. This format is useful to specify a list of initial constraints for the *Mfold* and *Kinefold* programs.

Bracket notation

It consists of two lines. The first line contains the sequence. The second line contains the set of associated pairings encoded by brackets and dots. A basepair between base "i" and "j" is represented by a "(" at position "i" and a ")" at position "j." Unpaired bases are represented by dots. The lack of *pseudoknots* in the secondary structure ensures that this notation defines a unique folding. This format is widely used in the Vienna Package. If no structure is detected then the message "No structure found" is displayed. The first explanation is that the sequences actually do not share a common structure. *Unfortunately*, there are other cases where caRNAc fails to infere correctly the structure. We shall see it in the next section.

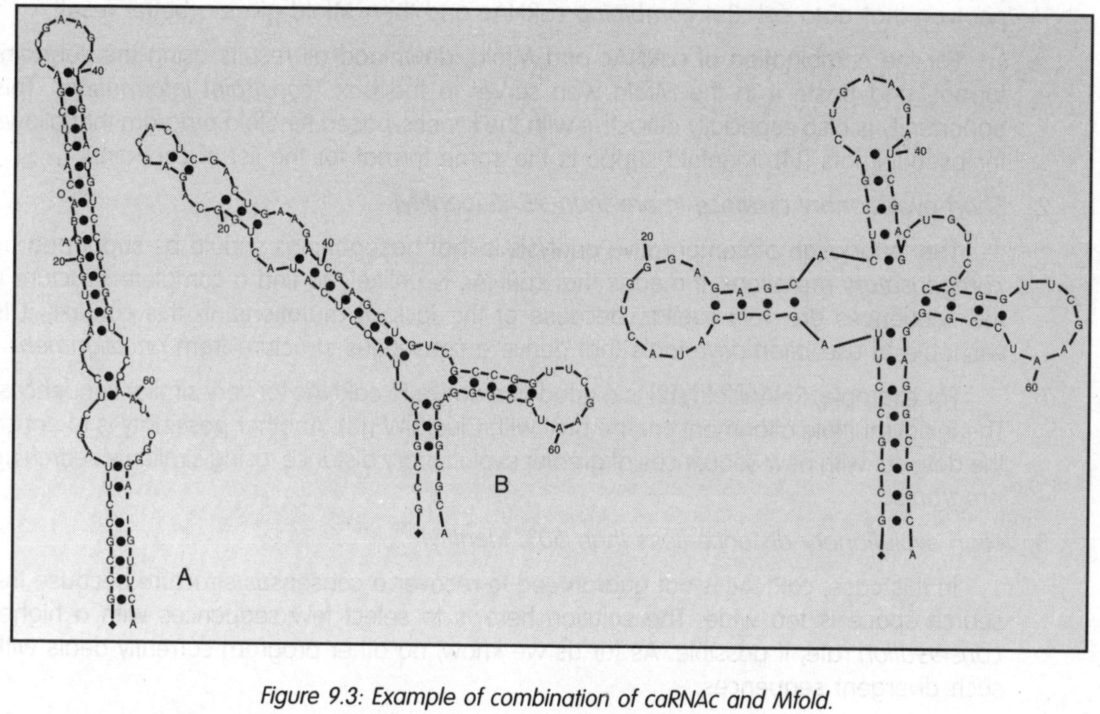

Figure 9.3: Example of combination of caRNAc and Mfold.

RNAfamily (button "Visualize all foldings with RNAfamily")

Allows the user to display all foldings at once. RNAfamily is a JAVA applet that is devoted to the *visualization* of *multiple* RNA sequences. It creates a plot using linear *backbone* representation. This is a concise *representation* that makes it *convenient* to compare several related structures at a glance. Each class of equivalent *helices* is assigned a color. RNAfamily includes the following *functionalities*: zooming, scrolling, selecting a stem, displaying the *nucleotidic* content.

It is also possible to download an archive storing all result files.

POINTS TO REMEMBER

We give here a list of pitfalls and limitations of the method. When possible, we suggest alternative programs that may prove to be more appropriate within the context. We also give further hints and rules of *thumbs* to *maximize* information from caRNAc output.

1. *Large unpaired regions*

   The philosophy of caRNAc is to privilege selectivity to sensibility. So it may happen that the prediction misses some stems. But these stems may be recovered afterward with external programs, such as Mfold. This is the case of the structure inferred for the three tRNA sequences (we choose this example on purpose).

   On the one hand, the basepairs inferred by caRNAc are globally correct, but there is obviously one *stem missing* to form the cloverleaf structure. This corresponds to the loop from position 45 to 65. One the other hand, the results obtained with Mfold alone are very poor on that data set. But *combining* caRNAc and then Mfold gives a better result.

   For the combination of caRNAc and Mfold, download all results using the contraint format, and paste it in the Mfold web server in the box "*constraint information.*" This opportunity is also especially attractive with the kynetic-based Kinefold program that allows for pseudoknots (14). Kinefold supports the same format for the list of contraints.

2. *Short evolutionary distance (more than 95 % identity)*

   The foundation of comparative analysis is that basepairings should be supported by *compensatory mutations.* It means that caRNAc is unlikely to find a complete structure if the sequences are very similar, because of the lack of mutations. In this context, it is wishable to use *alternative* tools that derive a consensus structure from an *alignment.*

   For example, RNAalifold (12) is a good alternative to caRNAc for very similar sequences. The initial multiple alignment can be built with ClustalW (13). Another possibility is to enrich the data set with new sequences at greater evolutionary distance, using similarity *searching programs.*

3. *High evolutionary distance (less than 50% identity)*

   In this case, caRNAc is not guaranteed to recover a consensus structure because the search space is too wide. The solution here is to select few sequences with a higher *conservation* rate, if possible. As far as we know, no other program currently deals with such divergent sequences.

4. *Pseudoknots*

The algorithm of caRNAc is not designed for handling pseudoknots. If sequence are short (less than 70 bases), it might be a major source of error. In this particular context, it is more advisable to use a comparative *pseudoknot*-friendly program, such as comRNA (15). Note that comRNA is a time and space consuming program compared to caRNAc. It is limited to smaller data sets. For longer sequences, *pseudoknots* are usually not a problem. Kinefold may be used afterward to complete the structure and identify potential *pseudoknots.*

5. *Alignment with the structures*

The multiple alignment tool allows the user to derive a structural alignment, taking into account both primary and secondary structures, from caRNAc output. Of course, the accuracy rate of caRNAc is not 100 %. Benchmark data show that *predicted* stems are usually *correct,* as soon as the number of stems is high enough to form a robust structure. In this context, some rare missing stems may be recovered afterward.

The situation is more complex with sparse structures containing mostly unpaired regions. It is a difficult task to decide if the stems *actually* exist or if they are false-positives occurring by chance. One solution is to compare the energy level of the sparse structure given by caRNAc with *randomized* equivalent data sets generated with shuffle-aln.pl If the free energy is significantly lower with the intial data set, sequences are likely to share a common structure. We plan to integrate this *functionality* in the web server in the very near future.

Chapter 10

Transposable Elements

In the past, a *bacterial species* has been defined in part by the G+C content of its genome, and it was generally assumed that the overall G+C content reflected the base *composition* throughout the genome-i.e., it was more or less constant from one gene to the next, indicating a common line of descent for the majority, if not all, of the genes on the replicon. Underlying this assumption was the general belief that *bacterial* genomes change slowly with time, and that the bulk of genes on a particular genome will have coevolved over a considerable period and will therefore have been subject to the same selective pressures.

In recent years, the sequences of several *bacterial replicons* (*chromosomes* and *plasmids*) have been reported (see also www.tigr.org/tdb/mdb/mdbcomplete.html). Analyses of these data have indicated that the *nucleotide* compositions of individual replicons are not homogeneous; rather, the data indicate a significant degree of genetic flux involving acquisition of DNA sequences from diverse sources at different times. This leads to the conclusion that *bacterial replicons,* as *gene collections,* are dynamic rather than static.

Over time, some genes are acquired and others are lost Those that confer selective advantage are preserved, while those that do not, atrophy and may be lost. Each replicon, whether it is a chromosome or a plasmid, appears to comprise a basic core that can be recognized because the DNA sequences have the same G+C content. In terms of the *bacterial* chromosome, this core encompasses the *housekeeping* genes that are essential for survival (i.e., growth and reproduction).

For *plasmids,* the only essential genes and sequences are those needed for replication. The genes of the core structure display common codon usage, as would be expected, given that they will have been subject to *prolonged coevolution.* Inserted into the core are sequences that differ in *nucleotide* composition both from the core and from each other. This variation is *indicative* of different evolutionary descents before the sequences were fused, and implies a flow of DNA between different *organisms* (i.e., horizontal gene transfer and recombination).

Lateral gene transfer involving mobile elements such as plasmids has been known for a long time; what had not been appreciated is the degree of sequence *reassortment* that has occurred among replicons, including *bacterial genomes,* in general. Among the genes in human pathogens that have been acquired in this way are those that confer *pathogenicity* and resistance to *antibacterial* agents. Although the sources of many of these *determinants* are not known, carriage of some exotoxin genes can be ascribed to a process termed *lysogenic conversion.*

This refers to latent infection of a bacterial cell by a bacteriophage (phage) that integrates into the chromosome and becomes dormant. An example of an exotoxin encoded by a lysogenic phage is the cholera toxin of *Vibrio cholerae* (%; others are the cytotoxin produced by diseasecausing strains of *Corynebacterium diphtheriae;* the Shiga toxins (1 and 2), produced by *Shigella dysenteriae;* and *streptococcal erythrogenic toxin.*

Hence, to understand gene flux it is necessary to know something about how DNA is transferred between bacterial cells and rearranged by recombination.

Gene Flow in Bacteria

DNA is transferred between bacterial cells by one of three mechanisms: conjugation, transformation, and transduction. Conjugation, mediated by plasmids and conjugative transposons, is a semi conservative replication process and normally involves transfer only of the element itself. The *persistence* of a plasmid will usually reflect the ability to replicate *autonomously* in the new host, while the persistence of a conjugative transposon will reflect the element's ability to integrate into a DNA molecule in the new host cell.

Although plasmid acquisition can expand the gene set of a cell, it does not by itself involve gene reassortment of the type detected by sequence analysis. Nonetheless, conjugation can potentially mediate transfer of any DNA sequence. Indeed, it was precisely the ability to promote transfer of chromosomal DNA between *Escherichia coli* cells that first revealed the existence of the sex factor F, a conjugative plasmid.

Transduction involves DNA transfer mediated by a phage. In this case a segment of DNA that is not part of the phage genome is incorporated into a phage particle, either in place of or as an extension of the genome. This *particle* then injects its DNA load into another bacterial cell. There, nonphage DNA sequences may be rescued by *recombination* into one of the replicons in the new host.

In contrast, transformation describes the genotypic alteration of a bacterial cell as a result of uptake of DNA from its environment. Some if not all bacteria can scavenge DNA released by others and rescue the genetic content. Again, incorporation of the foreign sequence requires *integration* into an existing replicon by some form of recombination. In both *transduction* and *transformation,* DNA sequences are usually rescued by RecA-dependent *homologous* recombination.

This can both remodel genes, as illustrated by the development of penicillin-resistant *Streptococcus pneumoniae,* and increase the gene pool, as seen in the transduction experiments of Stanisich et al.; homologous recombination is *exploited* in the gene knockout and allelic replacement strategies now *frequently* used in *molecular* genetics.

However, as the name indicates, homologous recombination requires the participation of DNA sequences that have *significant homology,* and the recombination commonly results in DNA replacement rather than addition. Additions to a replicon are usually the result of transposition.

TRANSLOCATION OF NUCLEOLIDE

The genetic event in which a DNA sequence is translocated from one site to another is called *transposition*. It is usually unrelated site on the same or a different DNA molecule. Transposition is, necessarily, a recombination event in the sense that DNA phosphodiester bonds are broken and reformed in the process.

Most transposition events involve genetic structures called transposable elements, which comprise discrete DNA sequences with defined ends that generally mediate their own transposition. Although the term *transposable element* implies a defined genetic entity, the term *transposition* indicates only that there has been a genetic rearrangement by DNA relocation, which can occur by one of several mechanisms.

Transposable elements are virtually ubiquitous in bacteria, and transposition is responsible for much of the DNA rearrangement seen among bacterial replicons. Genetic translocation may also involve DNA sequences not formally classified as transposable elements, such as pathogenicity islands and gene cassettes.

Model of Transposable Elements

Most reported DNA transpositions in bacteria involve transposable elements, either as the sole sequence transposed or as mediators of translocations of DNA sequences that are not normally part of a transposable element. Most *transposable* elements conform to a simple model, encoding a protein or protein complex called a transposase, that mediates *transposition*, and individual elements may or may not encode other functions.

The end sequences of the element are usually short inverted repeats (IRs) that are necessarily identical or nearly so, because they are binding sites for the *transposase*. The inverted configuration ensures transposase is positioned similarly at both ends of the element, i.e., facing outward, so it can cleave the *phosphodiester* bonds that anchor the element to its current carrier, both freeing it for transposition and guaranteeing that the element is conserved from one transposition to the next.

These terminal IRs are element specific and usually are only 15-40 by long. For a few *transposable* elements, the IRs are much longer (80-150 bp) and contain multiple transposase binding sites. In some cases, it has been shown that transposition involves formation of a protein-DNA complex called a transpososome in which the ends of the element are bound by *transposase* and *aligned*. Bringing the ends together *presumably* facilitates the subsequent insertion, where both ends have to be ligated essentially into the same site If the element encodes only the function needed for transposition and is 1-2 kb in size, it is called an insertion sequence (IS) element.

If the transposable element encodes functions other than those needed for transposition, e.g., a gene encoding antibiotic resistance, it is called a transposon. Such elements range in size from 3-4 kb (i.e., those that have only one or two additional genes) to elements more than 10 times this size, in some cases encompassing dozens of genes. In addition, certain phages, called transposing bacteriophages, use a form of transposition for *replication*. The best known of these is the *E. coli* mutator phage, Mu.

Related transposable elements encode related transposases and have related terminal IRs. However, not all *transposable* elements conform to the general description. Notable among these are the IS91 family of elements, which lack IRs, indicating that these elements use a different transposition mechanism from most others.

IS elements were the first transposable elements in bacteria to be recognized as such, although pride of place goes to phage Mu as the first bacterial *transposon* to be discovered. The concept of transposable elements, however, was proposed a decade or more before the discovery of Mu. The idea was first put forward by Barbara McClintock in the late 1940s, before the structure of DNA was known, to explain results obtained from her elegant and detailed genetic analysis of maize kernel *pigmentation*.

She suggested correctly that *variations* in maize kernel pigmentation were the consequence of gene activation/inactivation resulting from translocation of one or more *genetic* elements from one chromosomal site to another. *Physical evidence* to support the concept came first from bacterial systems in the late 1960s. These studies produced *incontrovertible* evidence for the existence of *small, discrete,* mobile DNA elements: IS1, IS2, and IS3.

The importance of transposable elements in bacteria was subsequently illustrated by the discovery that *clinically* important *antibiotic* resistance genes were associated with bacterial transposons. Transposons can be described as either composite transposons or complex transposons. The latter *designation* essentially indicates that the element is not a composite *transposon.* Composite transposons, for example Tn*5* and Tn*10* have modular structures with a central sequence that encodes the function(s) that *identifies* the element *phenotypically*, but *lacks* transposition functions.

This is flanked by two copies of the same IS element, forming either long terminal direct repeats or, more commonly, long terminal IRs. One or both of these IS elements provides the transposase for transposition, while the short IR sequences on the IS elements provide the sequences necessary for end *recognition.* When an IS element becomes a component of a composite transposon it does not immediately lose the ability to function as an independent element. Hence a composite *transposon* might *comprise* three distinct transposable sequencestwo long *terminal* repeats and the entire composite structure.

However, further genetic change and development of the composite element, a process that has been called *coherence,* can lead to fusion of the component parts and loss of the ability of one or both terminal IS elements to function *independently.* All that need be retained is a copy of the transposase gene(s) and the extreme short IR sequences. These minimal *requirements* have been exploited, for example, in the development of the mini-Tn5 series of transposons developed for *mutation* and *gene analysis.*

Complex transposons lack both a modular structure and, in general, long terminal IRs. These elements are usually delimited by short IRs, of 15-40 by depending on the element, as found on IS elements. The IRs flank genes that encode both the functions needed for *transposition* and those that identify the element by conferring a particular phenotype on the host cell. No part of the structure can transpose *independently* of the rest.

Some noncomposite transposons-e.g., Tn*3* and related elements-can form composite transposable structures analogous to IS-based composite transposons. Indeed, this is the basis of the in vivo cloning system that uses a deleted version of phage Mu, called mini-Mu, to mobilize *chromosomal genes* on to the broad host range plasmid RP4. Transposing phages a e typified by phage Mu, discovered in the early 1960s in a culture of *E. coli*. Phage Mu is a temperate phage, in that it can *lysogenize* its host, and is so named because Taylor correctly surmised that the increased numbers of *auxotrophic* mutants in populations of *E. coli lysogenized* with Mu arose as a consequence of its insertion into genes encoding *biosynthetic enzymes*. Transposing phages like Mu use a form of transposition to replicate, and differ from other transposable elements in that their lifecycles include an extracellular phase, in the form of phage particles.

Distribution of Transposable Elements

Transposable elements are widely distributed in nature among eubacteria and archeae, and are also found in lower and higher eukaryotes. IS elements, found in both *prokaryotes* and *eukaryotes*, are the most common, and hundreds have been identified in many different bacterial species, including Gram-positive and Gram-negative microbes and Archaebacteria, such as *Halobacterium* spp. and *Methanobrevibacter smithii*. Wherever a serious search has been made for transposable elements, examples have been found.

Indeed, the failure to find a transposable element is more likely to indicate a defect in the detection system than an absence of transposable elements. Furthermore, as the complete genome sequences of many different *microbes* are determined, new *putative* transposable elements are identified A.

However, determination of whether these elements are real and active, as distinct from being *pseudoelements* or *remnants* of obsolete elements, requires experimental evidence that they can transpose. Transposons have been found in many different *bacterial species*, often of clinical or *veterinary origin*.

Our interest in things that directly affect the human condition, especially health, probably explains the preponderance of known transposons, both composite and complex, that carry antibiotic resistance genes, and the relative paucity of transposons carrying other markers. Most bacterial transposons have been found on plasmids rather than on other DNA molecules, but this probably reflects the fact that once a gene is located on a plasmid, its ability to spread to other bacteria is greatly increased.

It also appears that, for some transposons such as Tn3, transposition to plasmids is much more frequent than transposition into the bacterial *chromosome*, despite transposition generally not being site specific and chromosomes being at least two orders of magnitude larger than most *plasmids*. The reason for this apparent ara ox is unknown.

Functional status of Transposable Elements

Insertion sequences and transposons are given the designations IS and Tn, respectively, and in each case, individual elements are identified by an italicized number (e.g., *IS10, Tn10)*. To avoid the confusion that might arise from different elements being given the same identification numbers by different groups of researchers, it was *suggested* and widely accepted that numbers should be allocated from a central directory.

This was administered by Esther Lederberg at the Department of *Medical Microbiology*, Stanford

University, California who periodically published lists of new assignments. As the number of elements discovered increased, it became clear that this simple numbering system was not sufficiently informative, and authors began to elaborate. Accordingly, additional letters were included to indicate the source of the element, and the numbering became specific to the species. For example, *ISM1* is an element from *M. smithii,* while *ISRm2* is an IS element from *Rhizobium meliloti,* and both are distinct from the elements.

IS*1* and *IS2,* respectively, which were the first transposable elements to be described. More recently, Chandler and Mahillion suggested that new IS elements be numbered as follows: IS followed by the first letter of the genus and then the first two letters of the species in which the element was discovered (namely, the convention used to name restriction endonucleases), plus a number chosen consecutively according to how many other new IS elements have been found in the species (~ Changes have also been instituted in transposon designations, in particular when numbering conjugative transposons), and specifically those from *Bacteroides* spp., because the conventional numbering system fails to indicate the nature of the element.

Such elements are now designated CTn plus a number. Nonetheless, the intent must remain to give each new transposable element its own distinct designation. A list of assigned IS numbers ca be, found at http://www-is.biotoul.fr.

Reporting Transposable Elements

IS elements, and in particular IS*1*, IS2, and IS3, were discovered as a result of studying strongly polar mutations in the galactose and lactose operons of *E. coli, i.e.,* mutations that prevented expression not only of the gene in which the mutation was located, but also of other genes in the operon located distal to (*downstream* from) the mutation. Such *mutations* represent a form of gene silencing.

Although these mutations behaved like point mutations in that they could be mapped to a single site and could revert spontaneously at low frequency, the frequency of true reversion was not enhanced by chemical treatments known to enhance the generation of base *substitution* and *frameshift* mutations. To account for this *anomaly,* it was suggested that the mutations were the result of insertion of small segments of DNA at the point off *mutation,* a concept first proposed by Barbara McClintock to explain some of the genetic *variation* in maize.

This idea was validated by a series of elegant genetic and biochemical experiments that used λ.gal-transducing phages and visualization of the IS element by electron *microscopy* of *heteroduplex* DNA. These experiments demonstrated the physical nature of transposable elements that, until then, had been simply a genetic concept. Further analysis established that these elements could also promote major DNA *rearrangements,* such as extended deletions and could serve as mobile *promoter* elements, able to activate the *expression* of a gene adjacent to the point of insertion. This early work has been ably summarized by StarlingerM).

VARIOUS CATEGORIES OF TRANSPOSABLE ELEMENTS

Bacterial IS elements are the simplest type of transposable element, and several hundred have been *discovered* and characterized to some extent, usually in terms of their sequences They rarely exceed 2 kb in size and may be as small as 0.5 kb, which limits their genetic capacity in most cases to one or two genes. Most, as already stated, have short-terminal IRs that define the extremities of the element, and many have a single open-reading frame (ORF) that is presumed to encode the transposase.

In a few cases, such as IS*1*, IS*3*, *IS10*, IS*50*, IS9*1*, and IS.*903*, this has been established experimentally by genetic and biochemical analysis. For many others, it is assumed that the putative genes encode transposition functions because the predicted gene products show significant homology to a known *transposase*. For putative IS elements discovered as a result of bacterial genome sequencing, the structural similarities to well-characterized IS elements are the sole evidence for IS status.

Some elements contain more than one ORF, with the smaller ORF overlapping the major one, wholly or in part. In one or two instances, the smaller ORFs have been shown to be real genes, encoding proteins that regulate *transposition* of the element, although in most cases this remains to be confirmed experimentally. Such regulation is desirable, given that *transposition* can be mutagenic and is therefore potentially lethal to the cell.

An interesting gene arrangement is found in IS*1*. At 768 bp, this is one of the smallest IS elements, and it has two short ORFs, *insA* and *insB'*, which are adjacent but in different reading frames. These sequences are cotranscribed, but while the former codes for a discrete peptide, the latter encodes the C-terminal segment of a fusion protein, InsAB', for which *insA* encodes the N-terminal segment.

To create this hybrid protein, a proportion of the ribosomes that initiate translation of the *insA* region of the transcript undergo a -1 frameshift toward the end of this region before translation is halted by the termination codon that signals the end of the InsA sequence. This frameshift is directed by the nucleotide sequence at the point of changeover, and switches the translating ribosomes to the reading frame used to accommodate *insB*.

Translation then continues uninterrupted through the *insB'* region of the transcript to the translational termination codon that indicates the end of InsB'. It is the fusion product, InsAB', that is the transposase for the element.

Both InsA and InsAB' are DNA-binding proteins and compete for the IRs of IS*1*. InsA competes with the transposase, denying it access to its sites of action and inhibiting transposition. InsA also represses expression of the insAB' transcript. The transposition frequency is determined by the relative activities of these two proteins and is dependent on the ratio of the activities rather than the absolute activity of either. The transposases of IS3 and like elements are also fusion proteins.

From sequence comparisons it would appear that most bacterial IS elements belong to one of a small number of extended families. Approximately 20 apparently unrelated familial groups have been identified, and members of most of these are found in diverse eubacteria and archaebacteria. The largest group comprises IS3-like and IS5-like elements, of which more than 70 of each are known; they are found in both eubacteria and archaebacteria, indicating, perhaps, an unusual degree of horizontal transfer and/or ancestral elements of considerable antiquity.

The most comprehensive database on IS elements currently contains information on more than 800 elements from more than 196 bacterial species, representing 523 individual ISs (http:/ /pc4.sisc.ucl.ac.be/ is.html).

Mechanism of Transposition

Most of our information on the mechanisms of transposition used by different types of IS element has come from studies of IS*1* IS*10* IS*50*, IS*91*, IS*903*, and *IS911*. An important early finding was that once IS element an ent has transposed to a new site, the element is bracketed by short direct repeats. Furthermore, while the same element at different sites of insertion is flanked by repeats of the same size, the sequences differ from one insertion to another.

These repeats, 2-12 by depending on the element, arise from *a* sequence found at the target site prior to insertion of the element, and hence involve DNA replication. It was quickly realized that this *arrangement* would result if, when processing the target site for transposition, the *DNA duplex* was cleaved at sites on each strand that are not directly opposite each other, but are displaced by a few base pairs.

Then, if the IS element, released from its donor site, is joined to the single-strand extensions created by the staggered cut at the target site, short, single-strand gaps are created at both ends of the element. DNA gap repair synthesis would then produce the same short sequence (i.e., the short direct repeats) at each end of the element. This is indeed what appears to happen with many transposable elements of all types-IS elements, composite and *noncomposite* transposons, and transposing phages.

Therefore, many different transposable elements create the same type of sequence arrangement by transposing, namely a sequence delimited by terminal inverted repeats located between short direct repeats. Hence, if such an arrangement is discovered when analyzing new bacterial DNA sequences, it is reasonable to infer the presence of a transposable element, particularly if there is a potential gene encoding a transposase-like protein bracketed by the terminal IRs.

Further experimental work is then needed to determine whether the putative element is still transpositionally active. As already pointed out, several examples of such discoveries have been reported as a result of bacterial genome sequence annotation. However, in the case of one or two IS elements, for example IS 91 and the related elements IS801 and IS 1294; transposition to a new site does not generate target site duplications, which may indicate either that the mechanism employs a flush cut at the target site, or that the element transposes by a mechanism fundamentally different from those used by most transposable elements, as do IS*91*-like elements.

The transposition mechanism of many IS elements, including IS*10* and 1S*50*, is conservative, in that the double-stranded form of the element is disconnected from the donor DNA molecule and then inserted, in toto, into the recipient DNA molecule. The simplest conservative mechanism is cut-and-paste, where the ends of the sequence to be transposed are brought together (synapsed) in a transposase-DNA complex. The transposase cuts both DNA strands at both ends of the element, releasing it from the donor DNA molecule, formally generating a free form of the element.

True free forms of transposable elements are not normally detected, and probably usually remain as a component of a transposition complex, the transpososome, which attacks the target site. The 3'-OH groups at both ends of the element generated on release from the donor molecule, serve as nucleophiles that are used to attack nonopposed 5'-phosphate groups on opposite DNA strands at the target site to initiate two transesterification reactions P31. These reactions require no additional energy input and connect the element to the target molecule by single-stranded

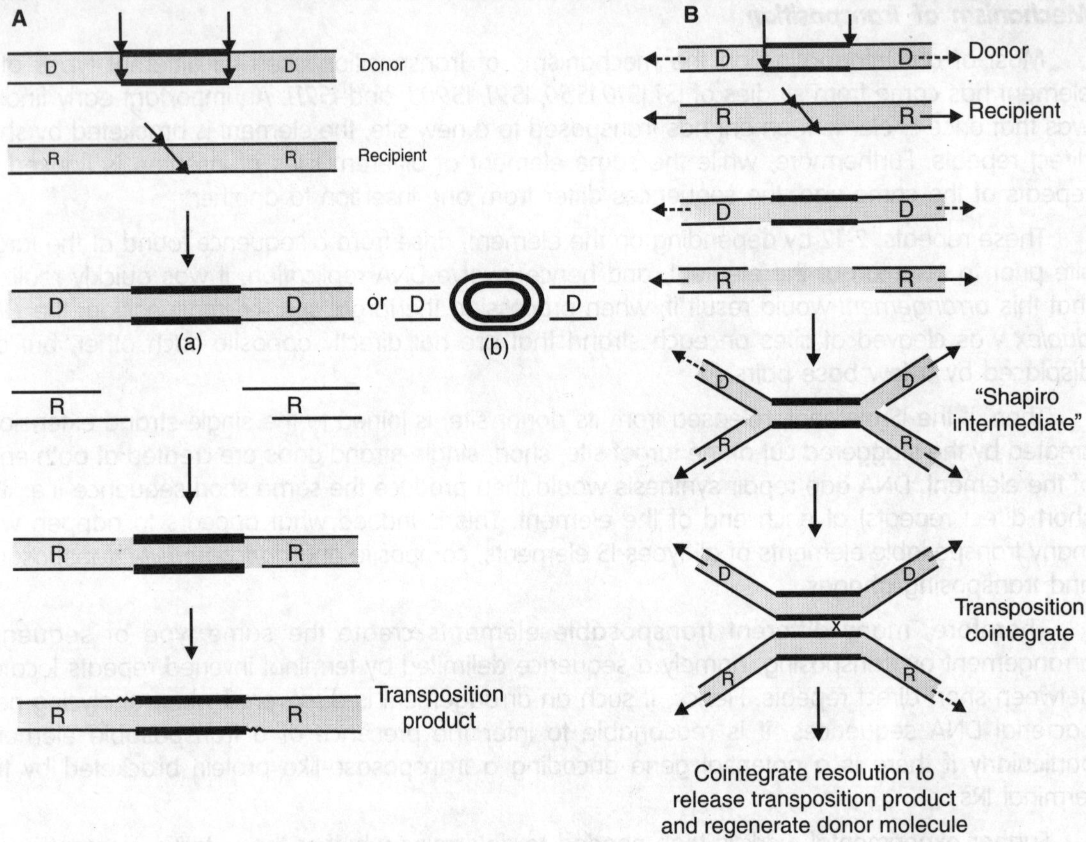

Figure 10.1: Showing the bacterial transposition. (A) Conservative transposition by "cut-and-paste" (B) replicative transposition via "Shapiro intermediate". (Fig. Contd.)

DNA. The short single-strand gaps are believed to be filled by gaprepair DNA synthesis, generating the short direct-target-sequence repeats that bracket the element.

The fate of the remainder of the donor DNA molecule, which is linearized when the IS element is excised, is unknown, but what little evidence there is suggests that it is probably degraded. Some IS elements transpose conservatively, using a mechanism that involves a true circular intermediate.

The element is disconnected from the donor molecule, and the ends of the element are joined to produce a circular form. IS3 and *IS911,* both of which are IS3 group elements, use such a mechanism. Production of the circular form of the element depends on two element-encoded proteins, OrfA and OrfAB, the latter being a fusion protein, the product of two open reading frames, *oif*A and *orf*B. As in the case of IS*1*, the two *orfs* are in different reading frames (0 and -1) and overlap to a small extent.

Again, production of OrfAB is the result of a directed ribosome frameshift. Both the OrfA and OrfAB proteins bind the terminal IRs of their cognate element, and both proteins are needed for high-frequency transposition (*cf*, the case of IS*1*, where the fusion protein InsAB is the transposase,

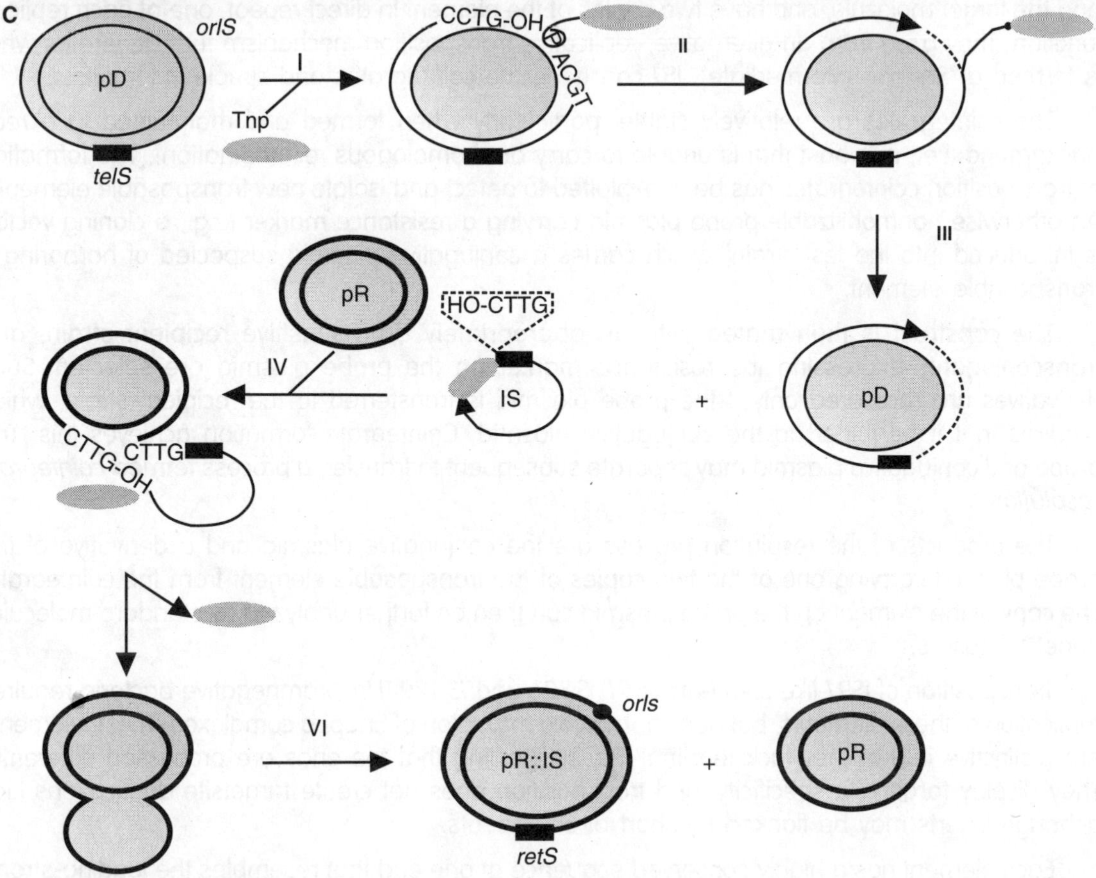

(Contd.) Figure 10.1: Showing the bacterial transposition. (C) RC transposition transposase proteins are boosted when needed for the second step of the transposition, namely insertion of IS911 *into a new site.*

while the second, smaller protein InsA inhibits transposition by competing with InsAB nonproductively for the IRs).

IS911 circles have been isolated and shown in vitro to be efficient substrates for transposition. When the element is disconnected from donor DNA and circularized, 3 by of the donor, linked to one end of the element, are also removed and form a spacer between the terminal IRs of the element. This is lost when the element is inserted into a new site. One consequence of circularization of *IS911* (and IS3) is the creation of a new promoter for expression of the transposase genes from the temporary structure.

This promoter is significantly stronger than that used to express the transposase genes on the linear structure. Circularization places the terminal IRs beside each other. The -35 box of the new promoter is on one IR, while the -10 box is on the other, with the additional 3 by ensuring correct spacing of the boxes in the new promoter.

Several IS elements that normally transpose in a conservative manner can also generate what are termed transposition cointegrates. These structures are fusions of the transposition donor and the target molecule, and have two copies of the element in direct repeat, one at each replicon junction. They arise from an alternative, replicative transposition mechanism that generates what is termed a "Shapiro intermediate". IS*1* can generate cointegrates and simple insertions.

The cointegrates are relatively stable, particularly when formed and maintained in *a recA* background (i.e., in a host that is unable to carry out homologous recombination). The formation of transposition cointegrates has been exploited to detect and isolate new transposable elements. An otherwise nonmobilizable probe plasmid carrying a resistance marker (e.g., a cloning vector) is introduced into the test strain, which carries a conjugative plasmid suspected of harboring a transposable element.

The construct is then mated with an appropriately drug-sensitive recipient strain, and transconjugants expressing the resistance marker on the probe plasmid are selected. Such derivatives are recovered only if the probe plasmid is transferred to the recipient strain, which requires that it be joined to the conjugative plasmid. Cointegrate formation achieves this. The probe and conjugative plasmid may separate subsequent to transfer, a process termed *cointegrate resolution.*

The products of the resolution process are the conjugative plasmid and a derivative of the probe plasmid carrying one of the two copies of the transposable element from the cointegrate. The copy of the element on the probe plasmid can then be further analyzed by standard molecular genetic analysis.

Transposition of IS*91*-like elements (IS*91*, IS*801*, and *IS 1294)* in Gramnegative bacteria requires replication of these elements, but does not involve formation of Shapiro complexes. These elements are distinctive in that they lack terminal IRs, suggesting that the ends are processed differently. They display target-site specificity, and transposition does not create targetsite duplications (.ice although inserts may be flanked by short direct repeats.

Each element has a highly conserved sequence at one end that resembles the leading-strand replication origins (Is-origin) of a number of plasmids found in Gram-positive bacteria that replicate by rolling circle (RC) replication. This mechanism generates a single-strand plasmid intermediate template, which is then converted to the more conventional double-stranded form. The other end of each element is also highly conserved, with a region of dyad symmetry preceding a tetranucleotide sequence identical to the preferred target site. The tetranucleotide delimits the end of the element.

The sequence arrangement at this end of the element resembles a Rho-inde-pendent transcription termination signal. Transposition inserts the IS element on the 3' side of the target sequence, so for the next transposition the same short sequence is present at both ends (one originating from the carrier DNA, the other from the element itself). However, whereas the old target site is separated from the element on transposition, the "target-site sequence" that defines the "trailing" end is retained, and the element remains intact.

Each IS*91*-like element has a single ORF that occupies most of the sequence and encodes a protein that clearly belongs to the family of RC-plasmid replication proteins (REPs) (W. This and other findings suggest that transposition of IS91 and related elements involves RC replication and, probably, a transient singlestranded DNA intermediate. Single copies of these elements can translocate markers adjacent to the trailing end of the element and can mediate replicon fusions.

Table 10.1: Composite Transposons of Some Bacteria.

Transposon	Size (kb)	Terminal elements	Target-site IS family	duplication	Marker(s)h
			Gram-negative bacteria		
Tn*5*	5.7	IS50 Inverted repeat	IS4	9	Km Bleomycin Sm
Tn*9*	2.5	IS1 Direct repeat	is*1*	9	Chloramphenicol
Tn10	9.3	IS 10 Inverted Repeat	IS4	9	Tetracycline
Tn*903*	3.1	IS903 Inverted Repeat	IS5	9	Kanamycin
Tn1681	4.7	IS1Inverted Repeat	IS*1*	9	Heat-stable toxin.
Tn2350	10.4	IS1 Direct repeat	IS*1*	9	Kanamycin
Tn2680	5.0	IS26[c]	IS6	8	Kanamycin
			Gram-positive bacteria		
Tn3851	5.2	Not *Reported*	Not *Reported*	Not *Reported*	Gentamicin Tobramycin Kanamycin
Tn4001	4.7	IS256 Inverted repeat	IS256	8	Gentamicin Tobramycin Kanamycin
Tn4003	3.6	IS257 Direct repeat	IS6	8	Trimethoprim

Composite Structure Transposon

As already discussed, a number of IS elements are able to function in pairs to mediate transposition of DNA segments caught between them. These modular structures, which have a copy of the same IS element at each end, are called composite transposons, and have been found in Gram-positive and Gram-negative bacteria. In most cases, the two copies of the IS element form terminal IRs, although in a few composite transposons they form terminal direct repeats.

The majority of known composite transposons encode resistance to one or more antibiotics. Such composite structures transpose because each is seen by the transposition mechanism simply as a "stretched" version of the IS element. The IS element is recognized by its transposase via the short-terminal IRs that define the limits of the element. The composite structure, necessarily delimited by the same short IRs, is essentially indistinguishable from the element itself, except that it is larger.

An interesting quirk of composite transposons is that, because transposition needs only a pair of IRs, when they are part of a circular molecule an event called "inside-out" transposition can occur. Instead of recognizing the outside IR of each terminal IS element of the transposon, the transposase recognizes the inside pair of IRs adjoining the central domain. Transposition using these IRs disconnects the IS elements from the central section that characterized the original transposon, which is replaced with the rest of the replicon.

Although composite transposons are conveniently considered as a group because of their common construction, they may not be phylogenetically related, and it is important to remember that just as individual IS elements may transpose by different mechanisms transposition mechanisms of composite transposons may also differ.

One recently discovered manifestation of composite transposon formation concerns genes encoding CTX-M β-lactamases, which are found on a number of plasmids in some cefotaxime/ceftazidime-resistant clinical isolates of species such as *Klebsiella pneumoniae*. These plasmid-associated extended spectrum β-lactamase genes have emerged throughout the world in recent years as threats to chemotherapeutic efficacy, and are thought to have originated from the chromosomes of *Kluyvera* spp.

They have been mobilized from *Kluvvera* chromosomes in the form of composite transposons by the IS*1380*-like element *ISEcpl*. Unusually, however, these composite elements contain only one copy of *ISEcpl* as a terminal sequence, which provides transposition function and the IR sequence at one end of the transposed sequence. The other terminal IR sequence needed for transposition originates not from a second copy of *ISEcp1*, but from a sequence that is found fortuitously on the *Kluvvera* chromosome, close to the $bla_{CTx\text{-}M}$ gene, where it is mistaken for an *ISEcp1 IR* sequence by the transposase.

Accordingly, the mobilized $bla_{CTx\text{-}M}$ gene is flanked by an intact copy of *ISEcp1* and a homolog of its terminal IR sequence, emphasizing the point that the necessary and sufficient components for transposition are a pair of appropriately oriented IR sequences and the cognate transposase. Some IS elements, notably IS*10*, IS*50*, and IS*903*, were discovered because they form the terminal repeats of composite transposons encoding antibiotic resistance.

IS*10* provides the long terminal IRs of Tn*10*, a transposon that encodes inducible resistance to tetracycline, whereas inverted copies of IS50 constitute the terminal repeats of Tn5, which has an operon encoding resistance to kanamycin/neomycin, bleomycin, and streptomycin, although the last is not expressed in some hosts, including *E. coli ('Resistance* to kanamycin/neomycin is also encoded by *Tn903,* in which the single resistance gene is flanked by inverted copies of *IS903*. *IS10* and IS50 both belong to the IS4 family of elements, while *IS903* is an IS5-like element.

It is likely that the gene for the SHV β-lactamase found on several R plasmids has been recruited on more than one occasion from the chromosome of *Klebsiella pneumoniae* on composite transposons generated by the IS6-like element IS26 (Matthew Avison, University of Bristol, personal communication). IS50 encodes two proteins, Tnp (transposase) and Inh (inhibitor of transposition). These proteins are virtually identical and differ only in that Tnp is 55 amino acids longer than Inh at the N terminus.

Tnp and Inh are both primary translation products, each being produced from its own transcript. In Tn*10,* transposition is controlled by regulating production of IS 10 Tnp by means of a 69-nt antisense RNA that anneals to the *tnp* mRNA to block translation. The *tnp* mRNA and antisense RNA are not equally stable; the former has a half-life of approx 40 s, whereas the antisense RNA has a half-life of approx 70 min.

The molar ratio of these RNA species in steady state is approx 20:1 in favor of the antisense RNA, which means that IS 10 Tnp is synthesized at its highest rate immediately after the element is transferred to a new cell (a case of zygotic induction), with synthesis being markedly reduced thereafter. In contrast, *IS903* limits transposition by producing an unstable Tnp that does not accumulate. Hence, the frequency of transposition is determined by the rate of transposase production at the time.

Transposons of complicated Genetic structure

A complex transposon has a more complicated genetic structure than an IS element or a composite transposon, as genes that have no transposition functions (e.g., antibiotic resistance genes) have been recruited into the body of the element, rather than captured by "block assembly." However, it should be remembered that, just as the term *composite transposon* denotes only that the element has a modular structure with terminal IS repeats, so *complex transposon* implies only that the transposon is neither a composite transposon nor a transposing phage.

Other complex Transposons

The classic complex transposon is Tn3, which is derived from resistance (R) plasmid R1. It is essentially the same as Tn*1*, which originates from another R plasmid, RP4, and was the first transposon to be discovered. Both elements were originally called TnA, and their transposition functions are fully interchangeable. These transposons encode TEM1 and TEM2 β-lactamases, respectively, that differ by only a single amino acid (a Gln37Lys substitution) and have virtually identical substrate specificities, conferring resistance to some β-lactam antibiotics, notably ampicillin, carbenicillin, and early cephalosporins.

Tnl and Tn3 are widely distributed among Gram-negative bacteria from clinical and veterinary sources, notably the Enterobacteriaceae and *Pseudomonas aeruginosa,* and are usually found on plasmids. Tn3 is more common. In the last few years, many variants (>100) have evolved from both enzymes, including the so-called extended-spectrum TEMs, which can hydrolyze third-

Table 10.2: Some Tn3-Like Transposons.

Transposon	Size (kb)	Terminal IRs (bp)	Target (bp)	Marker(s)ª	Subgroup
				Elements from Gram-negative bacteria	
Tn*1*	4.957	38/38	5	Ampicillin	Tn*3*
Tn*3*	4.957	38/38	5	Ampicillin	Tn*3*
Tn*21*	19.6	35/38	5	Mercuric ions Streptomycin Sulphonamide	Tn*501*
Tn*501*	8.2	35/38	5	Mercuric ions	Tn*501*
Tn *1000*	5.8	36/37	5	None	Tn*3*
Tn*1721*[b]	11.4	35/38	5	Tetracycline;	Tn*501*
Tn*1722*[b]	5.6	35/38	5	None	Tn*501*
Tn*2501*	6.3	45/48	5	None	Tn*3*
Tn3926	7.8	36/38	5	Mercuric ions	Tn*501*
				Elements from Gram-positive bacteria	
Tn*551*	5.3	35	5	Erythromycin	Tn*501*
Tn*4430*	4.2	38	5	None	Tn*4430*
Tn4556	6.8	38	5	None	Tn*4652*

generation cephalosporins such as ceftazidime and cefotaxime (which TEM 1 and TEM2 do not hydrolyze), and TEM variants that are insensitive to β-lactamase inhibitors, such as clavulanic acid, that efficiently inhibit the parent enzymes.

The speed with which these variants have become established owes much to the widespread distribution of the parent genes on Tn1 and Tn3, and also, perhaps, to the fact that acquisition of Tn3 can confer a growth advantage both in vitro and in vivo in the absence of antibiotic selection Tn3 is typical of an extended family of phylogenetically-related transposons from both Gram-negative and Gram-positive bacteria. Among them they carry genes encoding resistance to many antibiotics, to mercuric ions, and catabolism of nutrients.

These elements have different, but clearly similar, short (35-40 bp)-terminal IR sequences and related transposition genes. Each encodes a transposase, TnpA, of approx 1000 amino acids. Most also encode a second recombination enzyme, TnpR, that is a site-specific recombinase of the resolvase/DNA invertase family, which mediates the second step in a two-step transposition mechanism, a reaction called cointegrate resolution. TnpR requires a specific site for action, called *res,* located beside *tnpR.*

A few elements, typified by Tn*4430,* encode an alternative cointegrate resolution enzyme, Tnpl, that belongs to the phage λ integrase family. One member of the Tn3 family is only 3.2 kb long. IS*1071* looks like a typical IS element in that it consists only of the transposase gene and terminal IRs. The transposase is only distantly related to those encoded by other Tn3like elements.

Most Tn3-related transposons belong to one of two main branches of the family. On each branch the transposition functions of the transposons are more closely related to each other than they are to those of elements of the other branch. The parent element from which Tn3-like elements evolved is unknown, but it has been speculated to be *IS1071*. The relatively small (5 kb) element Tn3 is the type element for one branch that includes *Tn1000* (also known as y*S*), a cryptic element, and Tn*1546,* an element from *Enterococcus faecium* that encodes resistance to vancomycin.

The transposition functions on Tn3-like transposons occupy approx 4 kb, irrespective of the size of the transposon. Tn21, the type element for the other main branch, is some-what larger (20 kb), and is also a resistance transposon conferring resistance to streptomycin, spectinomycin, sulphonamides, and mercuric ions. This last feature is typical of Tn21-like elements (R1 The Tn21-like elements have essentially the same core structures and transposition functions, and differ primarily in the number and type of other resistance determinants they carry.

The variation in the antibiotic resistance gene composition of Tn21 and related transposons is believed to be of recent evolutionary origin, and reflects the activities of integrons and gene cassettes. In addition to the two main family branches, there are some elements that are Tn3-like, but that are not closely related to Tn3, to Tn21, or to each other.

There are two arrangements of transposition genes on Tn3-like transposons. On Tn3 and its close relatives, *tnpA* and *tnpR* are adjacent, but have opposite orientations and are transcribed divergently; *res,* the specific site for cointegrate resolution, is located in the intergenic region. On *Tn501* and closely related elements, these features are arranged in the order *res, tnpR, tnpA,* with the two genes similarly oriented and transcribed away from *res.* In both arrangements the *tnpA* gene is at one end of the element and runs into one of the terminal IRs of the transposon.

One consequence of the gene arrangement in Tn3 is that TnpR, by binding to *res,* functions as *a tnpA* repressor as well as a site-specific recombinase; Tn3 derivatives with null mutations in *tnpR* transpose at markedly elevated frequencies and create cointegrates that are inefficiently resolved in a RecA-proficient host, and are essentially stable in *a recA*-defective background.

The mechanism of transposition of Tn3-like elements involves semi-conservative replication and recombination and generates cointegrates via Shapiro complexes. Transpositions create targetsite duplications, generally of 5 bp, which result from the target site being opened by staggered cuts across the duplex DNA. The mechanism differs from conservative mechanisms in that the element is only partly disconnected from the donor site.

Single-strand cuts are made by the transposase at both ends of the element, on opposite strands of the duplex DNA, to expose 3'-OH groups. These nucleophiles are then used to attack phosphodiester bonds at the target site to execute two transesterification reactions on each strand of the DNA, 5 by apart. The product is a "Shapiro intermediate," a structure proposed by J. A. Shapiro (and independently, at the same time, by A. Arthur and D. Sherratt). In this intermediate, the transposon is joined to the target, but remains attached to the donor.

The transesterification reactions necessarily generate 3'-OH groups at both of the new junctions, which can be used to prime replication of the transposon by host-cell enzymes. When donor and target sites are on separate replicons, transposition results in replicon fusion producing a circular transposition cointegrate that contains the donor and recipient replicons joined by two copies of the transposon in direct repeat, one at each junction (Tig-1r).

The cointegrate is normally processed to release two replicons-one a derivative of the recipient DNA molecule carrying a copy of the transposon, and the other indistinguishable from the transposon donor. This process, termed *cointegrate resolution,* is mediated by TnpR (or Tnpl), the site-specific recombinase, using the pair of *res* sites on the cointegrate, one on each copy of the transposon. Cointegrate resolution can also be effected by host RecA-dependent recombination, but is less efficient.

This replicative mechanism of transposition requires both ends of the transposon, which are processed in the same way, explaining why the transposon terminals are perfect or near-perfect IRs. The job of the sequences is to bind transposase and direct the cleavage reactions precisely to the ends of the element, preserving both sequences as parts of the transposon in the process. It is surprising, therefore, to find that transposition can still occur when one end of the transposon has been deleted, albeit at a much lower frequency.

One-ended transposition is mediated by some but not all Tn3-like transposons, and requires one IR sequence and the appropriate transposase to generate what are essentially replicon fusions. The entire donor replicon, or part thereof, is inserted into the recipient DNA molecule. One consequence of this is that the products of this form of transposition, unlike those with the intact element, are not of uniform size. Inserts in different transposition recombinants form a nested series.

One end of all the inserts is the solitary IR sequence. Consistent with the mechanism being transposition, different insertions are flanked by different 5-bp direct repeats. The products are not generated by the replicative transposition mechanism described for Tn3-like elements, although components of that mechanism are clearly needed. Rather, the products have the hallmark of rolling-circle replication, as for IS91-like elements, where only one end is anchored.How the other end is determined in these aberrant reactions is not known.

Some Tn3-like transposons, including Tn3 but not Tn21, are subject to transposition immunity. In such cases, the native element will not transpose at a detectable frequency to another DNA molecule that already has a copy of the element. The signal for inhibition is a single copy of the element's terminal IR sequence.

For Tn3, transposition immunity can be largely eliminated by increasing the level of transposase in the cell by a knockout mutation in *tnpR.* Transposition immunity is not restricted to Tn3-like elements and has been reported for Tn7 and phage Mu, with the mechanism of transposition immunity in the latter having been well studied and characterized.

GENE CASSETTES

Tn21 and related elements form a subgroup of closely related, relatively large (approx 20 kb) transposons that confer resistance to many antibiotics, including b*-lactams, aminoglycosides,

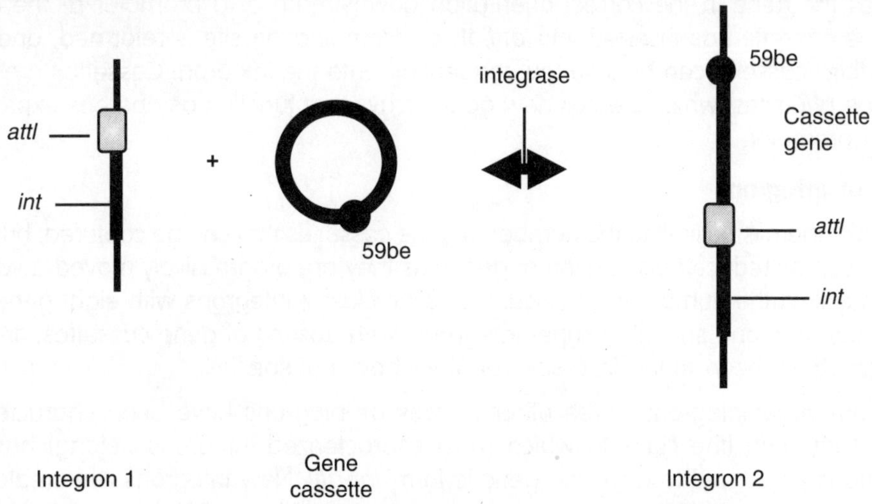

Figure.10.2. Capture and release of gene cassettes in Bacteria.

chloramphenicol, and trimethoprim. One characteristic of Tn21-like transposons is that antibiotic resistance genes cluster at essentially the same position in the different elements P40, indicating nonrandom insertion.

Given the variety off resistance genes involved (approx 40 to date), the fact that they migrate to the same point within an approx 20-kb sequence suggests that the different sequences have been captured by a site-specific recombination system. This involves elements called *integrons* and small, mobile, gene-sized DNA sequences called *gene cassettes.*

An integron is a genetic entity with a site used to capture other DNA sequences (the gene cassettes) by site-specific recombination. The recombinase, called *integrase,* is encoded by the *int* gene, which is adjacent to the recombination site *attl.* Since the initial discovery of integrons on Tn21-like transposons, several different integron classes, classified according to their integrase genes, have been reported. Tn21-like transposons carry class I integrons, each of which encodes essentially the same integrase.

Many of these integrons accommodate a gene for sulphonamide resistance, *sull,* and a truncated gene, *gacEdl,* which in its intact state encodes resistance to quaternary ammonium compounds. Both genes are believed to be remnants of gene cassettes that have suffered partial deletion, locking what is left of each cassette into the integron, so that *gacEdll sull* now forms one end of the integron.

Hence, the distinctive array of a class 1 integron is *int-attl-gene* cassette(s)-*gacEdl -cull.* A gene cassette is a genetic element that exists, formally at least, as a nonreplicating DNA circle when moving from one integron to another. It is usually a linear structure found in the chromosome, plasmids, or transposons such as Tn21. Gene cassettes are usually 500-1000 by in size, which is large enough to accommodate a single gene, often an antibiotic-resistance gene (MT).

In addition, each has what is called a 59 base element (59 be) or *attC* site that is paired with *attl* for integrase-mediated recombination. In general, cassette genes lack promoters, but insertion at *attI* places the gene in the correct orientation downstream of a promoter at the end of *int.* When a gene cassette has inserted into *attI,* th e recombination site is reformed, and a second and then a third cassette can be inserted sequentially into the integron. Cassettes are separated by their 59 be *attC* sites, which creates new gene arrays that function as operons expressed from the integron promoter.

Categories of integrons

In principle, there is no limit to the number of gene cassettes that can be captured, but functional constraints, such as reduced expression of genes as they are progressively moved away from the promoter, may limit the number in practice. However, class 1 integrons with eight gene cassettes have been reported, and so-called super-integrons, with dozens of gene cassettes, first reported in *V. cholerae,* have been found in a several other bacterial species.

Apart from super-integrons, three other classes of integrons have been characterized. The basic class 1 integron (the class to which most characterized integrons belong) has no gene cassettes and is presumed to have the genetic form *int-attI.* New integrons are created when a gene cassette(s) is inserted at *attI,* or when a gene cassette(s) is deleted from an existing array. Although cassette insertion occurs at *attI,* individual cassettes can be removed from any part of the array by integrase-mediated excision using the 59-be regions that flank the cassette gene.

An excised cassette can be reinserted into the residual integron or into another integron at the relevant *attI* site, or may simply be lost because it fails to reinsert and is unable to replicate autonomously. The movement of gene cassettes in and out of integrons is a random process, and whether a particular arrangement survives or is lost depends primarily on natural selection; when the gene combination is of selective advantage to the cell, it will be established in the microbial population by clonal expansion, at the expense of less fit systems.

The archetype class 2 integron is part of the transposable element Tn7 and includes genes conferring resistance to trimethoprim and streptomycin. The integrase shows less than 50% identity to the class 1 integrases and is nonfunctional owing to a stop codon within the reading frame. Only one example of a class 3 integron is known. While most of the genes discovered on gene cassettes in integrons of classes 1 to 3 are antibiotic resistance genes, it is likely that these represent only a small fraction of all possible cassette genes, reflecting only the keen interest of microbiologists in bacterial drug resistance.

In this respect, it is worth noting that while superintegrons also contain gene cassettes carrying antibiotic resistance determinants, most cassettes in these elements do not encode antibiotic resistance. Integrons and gene cassettes are undoubtedly responsible for much of the drug resistance shown today by clinical bacterial isolates, particularly by members of the *Enterobacteriaceae* and *P. aeruginosa.* These systems provide a genetic flexibility that has been exploited to the full in the evolution of bacterial drug resistance.

That integrons have been found on the bacterial chromosome and on many plasmids, often as part of a transposon, simply underscores this flexibility. It is of interest, in this context, that the genes for the IMP and VIM metallo-β-lactamases, which can protect *P. aeruginosa* and some other Gram-negative bacteria against carbapenems such as imipenem and meropenem, are commonly found as gene cassettes on class I integrons.

SPECIFIC TRANSPOSABLE ELEMENTS

Members of one class of complex transposable elements, conjugative transposons, have the ability to move, not only from one DNA site to another, but also from one bacterial cell to another. This reflects the fact that these elements transpose by a conservative mechanism that utilizes a free circular intermediate, similar to the transposition intermediates of some IS elements such as *IS911,* and that each element encodes a set of functions for cell-to-cell transfer.

First described in Gram-positive bacteria, elements of this type have also been found in Gram-negative anaerobes, such as *Bacteroides* spp. Elements of this type transpose and promote their own conjugal transfer from one cell to another. They are not plasmids because they lack replication functions and are replicated as part of the carrier replicon. However,. they resemble plasmids because they can exist as circular extrachromosomal elements that can be transferred from one cell to another by conjugation.

Conjugative transposons are a subset of a larger group of mobile genetic elements that also includes *integrative "plasmids"*and genomic islands) and that use site-specific recombination systems to integrate into and excise from the carrier replicons upon which they rely for replication. Integrative "plasmids," such as R391 and R997 were originally thought to be true plasmids because they support conjugal transfer of their resistance genes, but numerous attempts to isolate them as plasmid DNA failed.

This was because they exist primarily as elements integrated into the host cell chromosome and not as autonomous units. Self-transmissible genetic elements that must integrate into replicons to ensure replication, such as conjugative transposons, integrative "plasmids" and genomic islands, have been termed *integrative and conjugative elements,* or *ICES* . One notable ICE Is SXT, a mobile genetic element from *Vibrio clzolerae* that confers resistance to chloramphenicol, streptomycin, sulphamethoxazole and trimethoprim

New Type of Element

A new type of *Tn916* was found in a clinical isolate of E. *faecalis* and confers resistance to tetracycline. *Tn916* is 18.4 kb and is essentially identical to *Tn918,* Tn919, and Tn925, which were recovered independently from different clinical isolates. The related element *Tn1545* is 23.5 kb and has genes encoding resistance to kanamycin and erythromycin, as well as tetracycline.

These transposons promote their own transfer from one bacterial cell to another at frequencies of 10^{-9} to 10^{-5}. *Tn916* is a typical conjugative transposon. More than 50% of the structure is taken up by a block of conjugation functions at one end of the element. Conjugal transfer of the transposon is consolidated in the recipient by transposition of the element into the chromosome, where it is replicated as part of the chromosome. Transposition genes map to the other end of the element.

Knockout mutations in two genes, *xi s-Tn* and *int-Tn,* which are needed for transposon excision , prevent both transposition and conjugation. In contrast, conjugation functions are not needed for transposition, and elements with damaged transfer systems can still transpose from the chromosome onto a plasmid, although mutations in some conjugation genes do reduce the frequency of this type of event.

Transposition of Tn916-like elements occurs in two stages . First, staggered double-strand cuts are made at the junctions of the element with the current carrier DNA, generating a DNA fragment with a 5' single-stranded hexanucleotide overhang at each end. The ends are then joined to create a circular form of the element with a 6-bp heteroduplex spacer separating the ends.

This may then be reinserted into the host cell chromosome, be transposed onto a plasmid in the same cell, or be conjugally transferred to another cell and then transposed into the chromosome or a resident plasmid. The mechanism appears to be similar to the integration-excision system of phage k and the transposition proteins are related to the Int and Xis proteins used by phage k for integration into and excision from the *E. coli* chromosome.

TRANSPOSITION OF SITE-SPECIFIC ELEMENTS

Most transposable elements transpose into many sites on bacterial chromosomes or plasmids, although bias towards insertion into particular regions of replicons has been reported. For example, IS] and Tn3 appear to favor insertion into AT-rich tracts, and Tn5, which usually displays very low target-site specificity, has nonetheless been reported to have insertion hotspots.

This behavior reflects a bias on the part of the particular transposition system, but what governs the behavior is generally not known. In contrast, a few transposable elements have a marked tendency to insert in one particular site. Notable among these is the complex transposon Tn7.

Additional type of elements

Tn7, first identified on R plasmid R483 has two resistance genes, *dfil* and *aadA,* encoding dihydrofolate reductase I and aminoglycoside adenylyltransferase A, respectively, conferring resistance to trimethoprim, and to streptomycin and spectinomycin. These resistance genes comprise gene cassettes that are part of the class 2 integron carried on the transposon.

The integrase gene has sustained a nonsense mutation that severely truncates the integrase protein, inactivating it. Accordingly, movement of the gene cassettes requires complementation of the damaged integrase. The transposon is 14 kb in size, generates 5-bp target-site duplications on transposition, and inserts at high frequency in only one orientation into single sites on several bacterial chromosomes, including those of E. *coli, K. pneumoniae,* and *Pseudomonas* and *Vibrio* spp.

It also transposes onto plasmids with little or no site specificity, but at a markedly lower frequency. High-frequency, site-specific transposition occurs into a plasmid that has been engineered to have a copy of the chromosomal insertion site, which in *E. coli* is at 84', between *phoS* and *glmS*. The transposon inserts into the transcriptional terminator of *glmS,* about 30 nucleotides from the translation stop codon, which does not interfere with *glmS* operation.

The target site where Tn7 is inserted, which determines the 5-bp duplication, and the attachment site, *attTn7,* which is necessary for locus specificity, are not coincident. The *attTn7* sequence is approx 50 by and includes a small part of *glmS,* and although the *attTn7* sites in different bacterial species are homologs, the insertion sites are not, indicating that they are determined simply by their distance from *attTn7*.

Hence, Tn7 uses sequence data stored in a highly conserved gene, *glmS,* to identify a suitable insertion site where transposon insertion will cause little or no damage. Tn7 is also unusual in that it has five transposition genes, *tnsA, B, C, D,* and *E,* rather than the one or two seen with most other transposable elements. TnsA, B, and C are required for all transpositions. TnsAB is the transposase and is activated by TnsC. TnsD is needed for site-specific transposition (i.e., *attTn7* recognition), but is replaced by TnsE for transposition into randomly chosen sites 0-3?).

Whereas for other elements the transposition proteins are multifunctional, those of Tn7 have discrete roles. Nonetheless, they function as a DNA-protein complex. In vitro analysis has revealed that transposition of Tn7 is conservative, with the transposon being disconnected from its carrier by double-stranded cuts at both ends of the element, creating trinucleotide ssDNA overhangs, which come from the carrier DNA, and recessed 3' OH groups, which mark the boundaries of the element.

These nucleophiles are directed to attack the target, first causing breakage of phosphodiester bonds on both strands, with a 5-bp stagger, and then ligation of Tn7 into the gap via transesterification reactions. The resulting intermediate structure has the transposon flanked by very short single-strand gaps and 3-bp heteroduplexes formed from the trinucleotide overhangs attached to the transposon and one strand of the 5-bp target site.

It is presumed that the short stretches of mismatched sequences are removed and that DNA duplex continuity is restored by the gap repair mechanism of the cell. Site-specific Tn*7* transposition in vitro involves a DNA-protein complex of transposon donor DNA, target DNA, transposition proteins TnsA. B, C. and D, and ATP. TnsC is an ATP-binding protein. Cell-free transposition is authentic in the sense that Tn7 transposes into a target plasmid with *attTn7* in a site- and orientation-specific manner *03 1·33).*

The process is also subject to transposition immunity, as is Tn7 transposition in vivo (J·". While the terminal IRs of Tn*7* are 30 bp, sequences 75 by from one end and 150 by from the other are needed for transposition. Each terminal sequence contains multiple imperfect repeats of a 22-bp sequence that has been shown to bind TnsB. These are thought to be nucleation centers for the assembly of the transposition complex.

Element Encoding Resistance to Antibiotics

Other elements that show insertion specificity are Tn*502*, a 9.6-kb transposon found in Gram-negative bacteria, that encodes resistance to mercuric ions, and Tn*554,* a transposon from *Staphylococcus aureus* that encodes resistance to erythromycin and spectinomycin. Tn*502,* like Tn*7,* displays site and orientation specificity, but inserts into plasmid RP1 (RP4) rather than the bacterial chromosome.

When the insertion site is deleted, Tn*502* can transpose into many sites on the remainder of the plasmid at a significantly lower frequency, a behavior pattern that also resembles that of Tn*7.* Although it encodes mercury resistance, Tn*502* is not obviously related to Tn*3*-like transposons such as Tn*21* and Tn*501,* which also encode this resistance. Unfortunately, Tn*502*is still poorly characterized and there is insufficient information to say whether it is related to other site-specific elements.

Site Designated Element Tn554

Tn*554* transposes primarily into a site designated *att*Tn*554* on the *S. aureus* chromosome in one orientation only. Five genes have been identified on the element: two resistance genes, *ery* and *spc,* and a three-gene transposition operon, *tnsABC*. The products of transposition differ from those of most transposable elements in that a small part of the carrier DNA is transposed with the element.

These few additional base pairs are located at one end of the element when in the donor site, and are transferred to the other end of the element after transposition. Accordingly, a different short sequence is carried into the target site with each sequential transposition. The available data indicate a cut-and-paste mechanism via a circular intermediate, but further work is needed to consolidate this interpretation.

TRANSPOSABLE ELEMENTS

Transposable elements such as IS elements and transposons have no existence independent of the replicons that carry them, circular intermediates not-withstanding, since these are likely to be transient and remain complex bound. Some transposable elements, however, do have an independent state. For some phages, such as the *E. coli* temperate phage Mu, transposition is an essential aspect of their life cycles. Bacteriophage Mu lysogenizes *E. coli* by inserting into the bacterial chromosome in an essentially random manner, using a conservative cut-and-paste transposition strategy.

When induced to replicate, it does so by multiple rounds of replicative transposition, via Shapiro intermediates, into many sites on the chromosome. The former process was clearly established by exploiting the host-cell DNA methylation system. Phage particles were produced in *a dam*$^+$ host, where the phage DNA was fully methy-lated on both strands. These were then used to infect a dam-defective host in which there is no DNA methylation.

The newly integrated phage DNA was found to be fully methylated, indicating that transposition had occurred without replication, rather than being hemimethylated, as would have resulted from a replicative mechanism. The linear genome of Mu is 37 kb, but phage particles have DNA molecules of 39 kb, with the extra length being derived from the host. The phage genome is packaged by a "head-full" mechanism, starting from one end.

The end of the phage DNA that is specifically recognized and released by cutting the carrier DNA 150 by from the end is the one nearest the phage's replication/transposition genes. Approximately 39 kb is packed into the phage head, and the DNA is then cut once more to release a copy of the phage genome flanked by two small regions of host DNA, 150 by on one side and approx 2 kb on the other.

The genome structure of Mu is unremarkable, showing the usual array of genes for phage construction (i.e., for head proteins, tail proteins, and so on). In addition, it has two replication genes, A and B, located close to one end (diagrammatically usually placed on the left). Several years after its discovery, it was realized that Mu replicates by transposition, and subsequent investigations into Mu replication in vitro have been invaluable to our understanding of the form of replicative transposition that involves Shapiro intermediates.

Because replication of Mu, initially at least, involves intramolecular transposition events, gross molecular rearrangements result, which are most dramatically illustrated by the fragmentation of the chromosome into many smaller circles of DNA, each of which carries at least one copy of Mu. This kills the cell, and is an inevitable consequence of multiple rounds of indiscriminate intramolecular transposition. Unlike Tn3-like elements, which transpose by essentially the same mechanism, Mu lacks a cointegrate resolution system.

Indeed, such a system would be superfluous, given that the purpose of replication is simply to create a bank of phage genomes that are processed and packaged into mature phage particles. It is of no consequence that cointegrates are not deliberately resolved, as the phage genome is released by other means, and lack of cointegrate resolution is no bar to further rounds of transposition.

Mu was the first transposing bacteriophage to be discovered. D108, which closely resembles Mu, is the only other transposing coliphage to have been described. The two share 90% homology, with the principal regions of divergence being the ends of the genomes. One notable difference is at the 5'-ends of the A-genes, which results in poor complementation. Other transposing phages have been found in *Pseudomonas* spp., but although the sizes of their genomes are strikingly similar to that of Mu, there appears to be little or no homology with it. In *Vibrio* spp. the frequency of auxotrophs among certain phage lysogens is significantly raised over the normal frequency, findings reminiscent of the early results with E. *coli* and Mu.

GENOMIC DESCENDANTS

Analysis of bacterial genome data indicates that bacterial chromosomes are not homogeneous, i.e., sequences on a particular chromosome do not all derive from the same phylogenetic source. Punctuating the genome core, which accommodates the *housekeeping genes* that are essential for growth and reproduction (these display common codon usage indicative of coevolution), are sequences with different G+C contents (indicating different evolutionary descents).

These sequences, termed *genomic islands,* are most likely to have been acquired by horizontal transfer. A subset of genomic islands, first detected in human pathogens and subsequently in animal and plant pathogens, contains genes encoding virulence determinants; these are called pathogenicity islands (PAIs). The origins of some PAIs are rather more evident than others, in that they are clearly lysogenic phages (such as those that encode the cholera and shiga toxins of *V. cholerae* and *S. dysenteriae,* respectively and transposable elements.

PAIs accommodate genes encoding virulence determinants such as adherence factors, invasion factors, iron-uptake systems, protein-secretion apparatus, and bacterial toxins. They tend to be relatively large sequences of 10-200 kb and are often flanked by short direct repeats, suggesting involvement of transposable elements or site-specific recombination systems in their capture: they often house genes encoding proteins related to transposases and integrases, as well as IS elements or remnants of such elements. PAIs are defined by the following characteristics, although an individual PAI may not satisfy all of them:

- Presence in pathogenic strains and absence/sporadic presence in less pathogenic strains of the same or a related species.
- Occupation of large chromosomal regions (often > 30 kb) and carriage of virulence genes (often many).

- tRNA genes, direct repeats, and/or insertion sequences at boundaries; (sometimes cryptic) mobility genes and sequences (e.g., insertion sequences, integrase and transposase genes, plasmid origins of replication) present within the element.
- Different G+C content from the remainder of the chromosomal DNA of the host bacteria.
- Found as compact, distinct units.
- Instability (prone to deletion).

A few PAIs are mobile, and in some instances movement is undoubtedly mediated by phage; examples would include the PAIs of *S. aureus* and *V. cholerae.* It is the possession of particular PAIs that distinguishes uropathogenic, enteropathogenic, and enterohemorrhagic strains of E. *coli* from each other and from avirulent ones PAIs are often located next to a tRNA gene, and the boundaries of the element may be marked by short direct repeats indicative of site-specific recombination.

The presence of integrase genes on PAIs also points to acquisition by site-specific recombination, mediated perhaps by phage-like integrases. In some cases there is experimental evidence to support this view. Other genomic islands harbor antibiotic resistance genes. One important example is the methicillin resistance gene, *mecA,* in methicillin-resistant *S. aureus* (MRSA). It encodes a new penicillin-binding protein (PBP2') that shows considerably less affinity for most -lactams than the normal complement of PBPs of *S. aureus.*

The *mecA* gene is found in genomic islands of 21-67 kb, designated *SCCmec* (staphylococcal cassette chromosome *mec)* elements that are inserted into specific sites on the *S. aureus* chromosome.

Insertion and excision of these genomic islands is mediated by a site-specific recombination system encoded by two genes, *ccrA* and *ccrB,* closely linked to *mecA*. A mobile genetic element, closely related to the *SCCmec* elements, but lacking *mecA,* has been identified in a methicillin susceptible strain, *Staphylococcus hominis* GIFU12263. Designated $SCC_{12263,}$ this element is a possible precursor of *SCCmec* elements and may serve as a vehicle for transfer of various genetic markers between staph lococcal species.

INSERTION OF ONE DNA SEQUENCE INTO ANOTHER DNA

The direct consequence of transposition is the insertion of one DNA sequence into another, which may disrupt a gene at the point of insertion, causing loss of gene function. Indeed, it was such a consequence that led to the discovery of transposable elements in bacteria. The mutagenic activities of transposable elements have been successfully exploited in bacterial genetic analyses to locate genes of interest.

A number of transposable elements have been used, and several have been engineered with considerable ingenuity to provide genetic tools that can be employed not only to discover genes of interest, but also to analyze their expression and to study gene product structure. Notable among these is the mini-Tn*5* family of synthetic transposons and the mini-*Mu* system, devised to enable in-vivo cloning.

Several ingenious "suicide systems" have also been devised to deliver transposons into cells, where they are established only if they transpose onto a resident DNA molecule, because the delivery vector itself fails to replicate after transfer. Since such transposition events are essentially random with respect to choice of target site, insertions may occur in many different genes and at many different sites within the same gene. The difficulty in practice is often that of identifying the desired insertion(s) among the many that are of little or no interest.

Transposable elements mediate DNA rearrangements other than straightforward self-insertion. A site at which an active transposable element is located may be a hotspot for deletions. These start precisely at one end of the element (either end is possible, but only one at a time) and extend for varying distances into the adjacent DNA. Inversions of DNA sequences adjacent to the element may also occur.

Such rearrangements are usually accompanied by duplication of the transposable element on the DNA molecule. These events are random and infrequent, and the rearrangements reflect further rounds of transposition of the element into adjacent sites on the same DNA molecule. This ability of transposable elements to cause DNA rearrangement was detected early in the studies of the first bacterial transposable elements to be discovered.

For elements such as Tn3 and bacteriophage Mu, which transpose via Shapiro intermediates, intramolecular transposition automatically generates deletions and inversions, dependent only on how the 3'-OH ends of the element attack the linked target. If each initiates nucleophilic attack on a phosphodiester bond on the same strand as itself, then the transposition event will result in deletion of the DNA sequence between the "old" and "new" sites and the newly created copy of the transposon.

The deleted DNA is released as a nonreplicating circular DNA species in which the ends of the lost DNA are joined to the copy of the transposon created by the transposition. In contrast, if each 3'-OH group attacks a phosphate group on the opposite DNA strand, then the transposition will create a duplicate but inverted copy of the transposon on the DNA molecule, and the two sections of the carrier DNA separated by the duplicate copies of the element will be inverted with respect to each other.

Many other transposable elements, including those thought to transpose by conservative cut-and-paste mechanisms, generate adjacent deletions and inversions. Precisely how this is achieved is poorly understood. In the case of some IS elements, some of these events may reflect homologous recombination across IS element duplications; single crossover events involving two directly repeated elements will release the intervening DNA sequence together with one copy equivalent of the element, in much the same way that bacteriophage k is excised from the *E. coli* chromosome to initiate phage replication by site-specific recombination.

If, on the other hand, the recombination were to involve inverted copies of the element, then the DNA sequence between them will be inverted. An analogy for this would be the invertible segments of phages Mu and P1 that control host range, and the mechanism of flagella phase variation in *Salmonella* Typhimurium although once again these particular DNA inversions involve site-specific recombination systems and not homologous recombination.

Hence, duplicate copies of a transposable element on a DNA molecule are loci of intrinsic instability and DNA rearrangement. Transposable elements can also act as fusion points for DNA molecules. If two circular DNA molecules each carry a copy of the same transposable element, then homologous recombination between them can result in molecular fusion to create a single circular molecule essentially identical to a transposiion cointegrate structure, i.e., one in which the two replicons are joined via a directly repeated copy of the element at each replicon junction.

The first resistance plasmids isolated, typified by R1 and NR1 have such structures; however, it is impossible to determine retrospectively how these constructions were formed, i.e., by recombination between copies of the same element on independent replicons or by a transposition event. Suffice it to sly. the arrangement does permit specific fragmentation of the molecules. Replicon fusion, be it the result of a replicative transposition mechanism or of homologous recombination, offers the opportunity for conduction, that is, conjugal transfer of a nonconjugative DNA molecule by virtue of its attachment, possibly temporarily, to a conjugative element.

Both sequences are transferred to the recipient as parts of a single DNA molecule, which may then fragment to release either the original two molecules (when the fusion involved recombination across homologous sequences) or one original and a derivative of the second now carrying a copy of the transposable element responsible for the fusion (when fusion reflected formation of a transposition cointegrate). Conduction systems have been used to study cryptic transposable elements, notably IS elements, using linked plasmid markers to follow the DNA rearrangements, e.g., IS91.

SUMMARY

The ability of transposable elements to insert into and to mediate gross molecular rearrangements of DNA can account for much of the macromolecular rearrangement that has been discovered among related bacterial plasmids, and provide the mechanism whereby some plasmids are able to integrate into other replicons, as when the F factor (and other conjugative plasmids) inserts into the chromosome of *E. coli* to create an Hfr strain.

Bacterial chromosomes are subject to the same rearrangements by the same means. It is now clear that many if not all bacterial chromosomes are molecular hybrids constructed from DNA segments that derive from diverse phylogenetic sources, giving these structures a mosaic appearance. Further, chromosomes are dynamic structures that are subject to continuous assault by all the mechanisms, physical and genetic, that can alter DNA molecules.

Nonetheless, chromosomal rearrangement appears to be less frequent than the abundance of transposable elements in bacteria would lead us to expect. Indeed, if chromosome rearrangement were significantly more common than has been found, then it is likely that bacterial species boundaries would be somewhat more difficult to define than they are.

However, this doesn't necessarily mean that the frequencies at which such rearrangements occur is less than expected; rather it may simply reflect a somewhat low incidence of survival of the products of such rearrangements. One possible reason for this is that many such rearrangements are either immediately lethal to the cell, because they involve loss of or damage to a vital function, or because the derivatives are less fit for survival in competition with their progenitor, so they fail to establish in the microbial flora.

Survival fitness reflects not only the complement of genes possessed by the microbe but also the temporal expression of those genes and the balance of activities of the gene products. It is reasonable to believe that, over the prolonged period of time that constitutes the evolutionary history of a microbe, the balance of gene expression has been honed to optimize reproductive capability and survival characteristics in the environmental niche(s) occupied by the microbe.

Accordingly, major genetic change is likely to upset this balance of gene expression to the detriment of the microbe. Although the change may appear small, the effect may be insidious, blunting the microbe's ability to compete in a variety of situations. Large deletion and inversion rearrangements are likely to be disadvantageous to the microbe in a competitive situation, while the consequences of gene acquisition by insertion will depend on the precise location of insertion and the balance of advantage/disadvantage imposed by the additional genetic information.

New sequences offer the opportunity of new growth potential; e.g., acquisition of antibiotic resistance genes allows the microbe to survive, perhaps even thrive, in situations where the antibiotic sensitive progenitor would perish; however, large-scale gene acquisition may not be without risk to the microbe, simply because the optimum balance of gene expression is inevitably altered by the introduction of the new genes. If they do not confer any immediate advantage, the nascent microbe may not be sufficiently competitive to survive long enough to exploit its new acquisitions.

This aspect of microbial physiology/bacterial genetics deserves more attention than it has received to date. How the environmental fitness of a microbe is determined is poorly understood, yet it is clearly a question of major importance in relation to microbes that impact upon human health and well-being.

Chapter 11

Mobile Elements

The most *successful* mobile elements are called *Alu* elements found in any genome. They have reached a copy number in *excess* of one million *copies*, making up more than 10% of the human genome. The level of amplification required to reach this high copy number has *created* an *enormous* number of *insertion mutations* resulting in human disease and genome evolution.

They also add extensive diversity to the genome by introducing alternative splicing and editing to a wide range of RNA transcripts. In addition, after insertion *Alu* elements contribute to a high level of genetic instability through recombination. This *instability* contributes to a *significant* number of germ-line *mutations* and may be an even bigger factor in cancer and/or aging.

ALU ELEMENTS

Alu elements have reached a copy number in excess of 1×10^6, representing more than 10% of the human genome. They are widely distributed across the entire genome, with only a relatively few regions that have few of them. *Alu* elements tend to be enriched in the GC-rich. gene-rich regions, with many Alu elements located in the introns of genes. The current rate of *Alu* amplification has been estimated to be in the range of one new insertion per 20-200 human births.

The *Alu* insertion process has the potential to damage the genome both through insertional mutagenesis, and through facilitation of unequal, homologous recombination events (2). Insertional mutagenesis by *Alu* causes approx 0.1 % of human genetic disease, whereas *recombination* processes contribute to a much higher level. Little is known about the influence of *Alu* insertions in somatic or transformed cells. They have the potential to contribute to genetic instability in *tumorigenesis, aging,* and other *somatic* disorders.

INTRODUCING TECHNIQUE

Understanding the mechanism of *Alu* insertion is critical to understanding how they influence genetic instability. A typical *Alu* element, shown schematically in Figure elsewhere in this chapter, contains an internal, RNA *polymerase* III promoter that allows *transcription* through the element

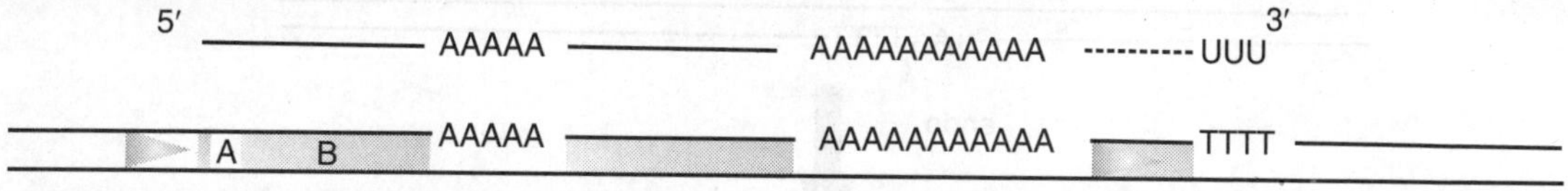

Figure 11.1: Structure of an Alu element. A typical Alu element is schematized with the arrowheads representing the short, direct repeats formed at the point of insertion. The left half of Alu contains the A and B boxes of the RNA polymerase III promoter. The two halves of the Alu dimer are separated by an A-rich region and the 3′ end of the Alu contain a variable, A-rich region. The line above represents a typical Alu transcript, which terminates in the sownstream flanking sequence at a terminator that generally contains four or more T residues in a row. The dotted portion of the line represents sequences downstream of the Alu element that will be variable between transcripts from different genomic loci.

and into the flanking sequence. *Alu* elements have a dimer structure, with both halves having an ancestral *derivation* from the 7SL RNA gene and have no capacity to code for proteins.

At the 3' end of each *Alu* element is an A-rich sequence of variable length and the elements are generally *flanked* by short, direct repeats of 5-20 by in length that are formed from *duplications* of the target site during *insertion.* It has been shown that the 3' A-rich region is routinely *lengthened* by the insertion process and rapidly shrinks after insertion.

It is thought that the RNA polymerase III-transcribed *Alu* RNA must associate in some way with at least the ORF2 product of L1 , which codes for *endonuclease* and reverse *transcriptase* activities. Thus, *Alu* is considered a *non-autonomous* mobile element, which is dependent on the *autonomous* L1 elements.

The endonuclease then cleaves the genomic DNA at a loose consensus sequence, and the genomic site primes reverse transcription of the *Alu* RNA, using the 3' oligo-A region of the *Alu* RNA as a template.

It has been hypothesized that *Alu* elements with longer A-tails are more active in the *re-trotransposition* process The typical integration is then finished off by the formation of a second-strand nick and integration of the end of the cDNA. However, significant portions of the events appear to complete the integration process by recombining with another *upstream Alu* element. This would cause sequences between the point of insertion and the upstream *Alu* element to be deleted.

It also seems likely that many *Alu* elements may not complete the insertion process. Thus, the initiation of the insertion may create genornic nicks that contribute to mutation and recombination processes in the cell, but show no evidence of *Alu* insertion. Although *Alu* amplification is clearly dependent on L 1 elements for their retroposition, there are definite differences in their amplification process. The most notable difference is that, using a tagged *Alu* reporter system, *Alu* insertion was found to not require exogenous L1 ORF1 expression.

PHYLOGENY OF ALU ELEMENTS

Alu elements began to amplify early in primate evolution. A precursor to Alu elements may have predated the primate/rodent divergence. *Rodent* have B1 elements that are also derived from the 7SL RNA gene. However, the B1 elements have a monomer structure, whereas the Alu elements have a dimer structure that formed early in primate evolution. Alu elements are found

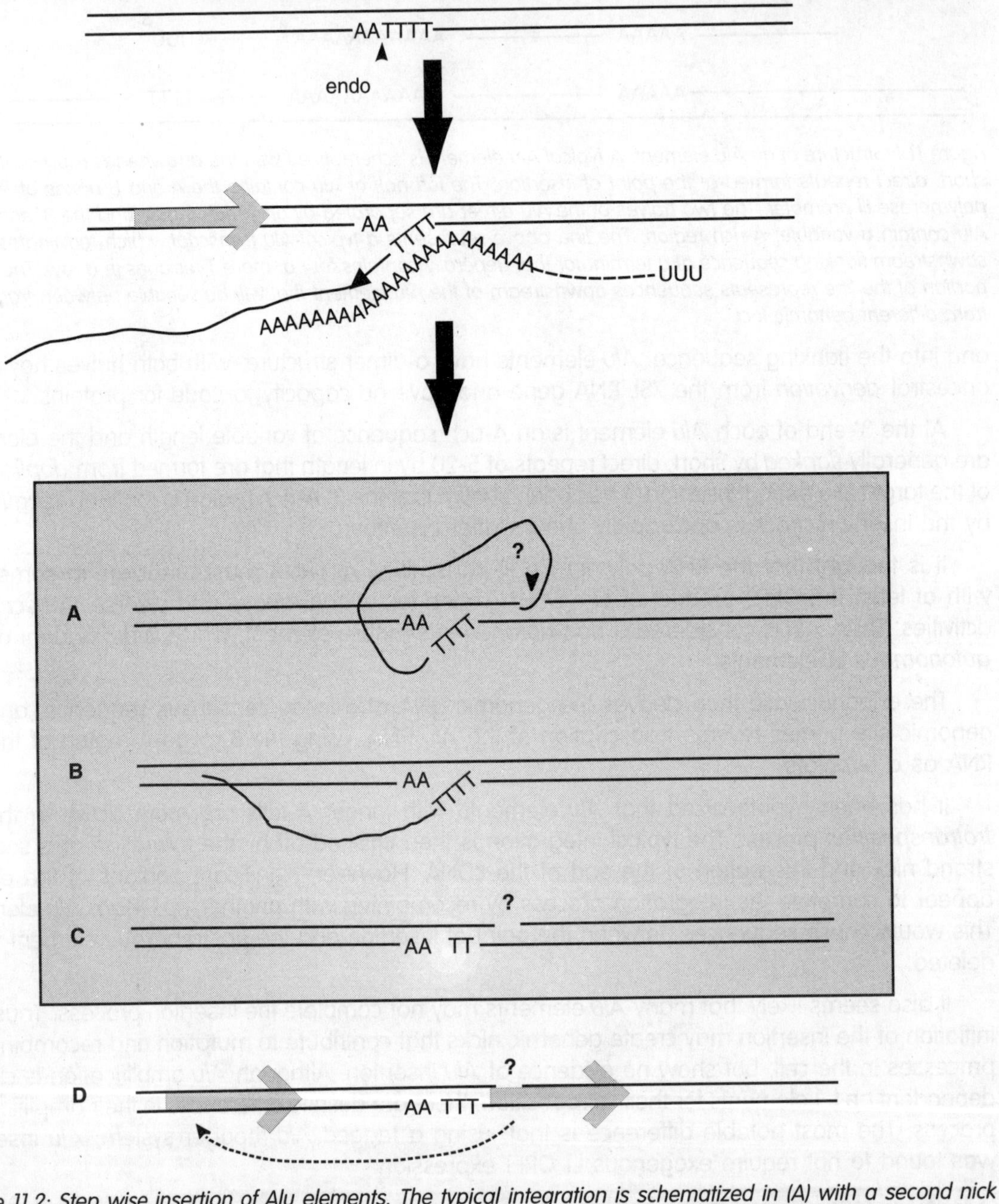

Figure 11.2: Step wise insertion of Alu elements. The typical integration is schematized in (A) with a second nick occurring from an unknown source, probably allowing integration of the 3' end of the cDNA using some sort of microhomology-driven priming. (B) An alternative mechanism that occurs occasionally where the 5' end finds a homology with another Alu element upstream. (C) It is possible that the integration process may abort prematurely. possibly with the intervention of DNA repair processes, leading to DNA damage. (D) Alu elements may undergo unequal recombination with other Alu elements nearby. We hypothesize that nicking by L1 endonuclease at consensus sites adjacent to Alu elements may facilitate this recombination process.

even in the prosimian primates. Alu elements began to *accumulate* in primates about 65 million years ago.

Because there is no specific mechanism for removal of *Alu* elements, their copy number has *continued* to increase throughout primate evolution. As the copy number increased, the sequence of Alu elements has evolved as well. Figure anywhere else in this chapter shows a *schematic* of Alu insertions during primate evolution. There were very high levels of Alu retroposition early in primate evolution. The current rate in human is nearly 100-fold lower than at the peak of *Alu* insertion.

At different stages of primate evolution, the *Alu* elements that amplified had distinct sequence differences that allowed them to be classified into *subfamilies.* Recent studies suggest that there has been a modest increase of Alu insertions in the human *lineage* relative to the other great apes The most likely *explanation* for the formation of different Alu subfamilies at different evolutionary times is that there are extremely few "*active*" Alu elements at any one time.

This allows the sequence of active elements to drift with time. Various *hypotheses* have been proposed for the features that limit Alu element activity. These include the flanking sequences of *Alu* elements influencing transcription rates and subfamily-specific changes interacting with the retrotransposition machinery. Currently, we favour the length of the A-tail as the most important factor, The A-tail grows on insertion but *shrinks* rapidly in evolutionary terms, *potentially silencing* individual *Alus.*

In addition, older Alu elements accumulate mutations that lessen their expression, as well as potentially their interaction with the retroposition proteins. Thus, only progeny from the most recent insertions are likely to maintain activity. Because the most recent inserts are also the most likely to be active elements, many of the most active elements will be *polymorphic* in the human population. Thus, different individuals in a population may be more or less prone to Alu *amplification* and it is the entire population that serves as a reservoir for *potent* active elements.

Alu Elements in Human Genome

Although Alu elements are spread throughout the human genome, they show some regions of higher density. In particular, Alu elements appear to be *preferentially* located in GC-rich genomic "isochores, while L1 elements are located in AT-rich isochores. However, most recent data suggest that *Alu* elements insert relatively randomly with respect to GC content and are selectively lost from the AT-rich regions, creating the bias seen.

Thus, *Alu* elements are likely to have a fairly random insertional mutagenesis potential, but are more likely to contribute to postinsertional recombination events in the GC- and generich regions of the chromosomes.

Genomic Insertion in Coding Area

Alu elements create an approx. 300-bp insertion mutation at any new genomic insertion site. If *Alu* insertion occurs in a coding exon, or near a splice junction, they are likely to disrupt the *appropriate* expression of a gene). Table elsewhere in this chapter shows a collection of most of the known cases of human diseases, caused fairly equally by these two types of *insertions.*

Although these represent a number of different diseases, reflecting the broad distribution of

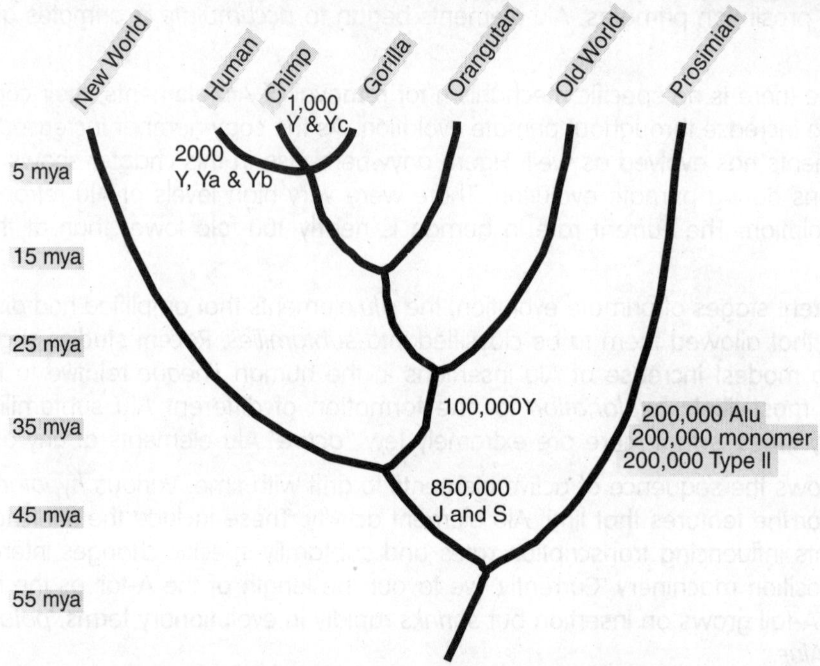

Figure 11.3: Phylogeny of Alu elements in the primates.

Alu elements throughout the human genome, there are multiple independent cases in four different loci. This is probably largely a case of ascertainment bias because those genes have been very heavily studied. However, there is also a strong preference seen for diseases associated with the X chromosome (10/23 cases).

There is much less of a bias for the X chromosome than has been seen for L1 element insertions, and probably also largely represents an ascertainment bias in favour of sex-linked diseases. In a few rare cases, *Alu* elements have inserted into an *exon* coding for the carboxy-terminus of a protein and has not *destroyed* the function of the protein.

In these cases, the *Alu* element has *incorporated* a portion of its sequence into the coding region of the protein, likely because that *portion* of the *protein* was not critical for function, but novel proteins can evolve this way.

Insertions in the middle of introns or between genes appear to be well-tolerated and to have minimal effects on genes. Many genes have introns that include dozens of *Alu* elements that make up more than half of their DNA. However, as discussed next, many of these initially harmless *Alu* insertions can become deleterious by contributing to recombination events, or through further mutation *altering* their properties.

As illustrated in Figure elsewhere in this chapter, it is possible that *Alu* elements sometimes begin the *insertion* process, but that cellular repair processes remove the initial primed element. This would result in a nicked DNA that could serve as a free end in a *recombination* event, as well as lead to mistakes in DNA repair and other DNA damage.

Such events would be the result of an attempted Alu insertion, but would leave no evidence

Table 11.1: Human Abnormalities due to Alu Insertions.

Actual Site	Chromosome	Abnormalities
3X HEMB (IX)	X	Hemophilia B
2X HEMA (VIII)	X	Hemophilia A
2X CLCN5	X	Dent disease
BTK	X	X-linked y-globulinaemia
IL2RG	X	XSCID
GK	X	Glycerol kinase deficiency
BCHE	3	Cholinesterase deficiency
CASR	3	Hypocalciuric hypercalcemia and hyperparathyroidism
MLV12	5	Leukemia
APC	5	Hereditary desmoid disease
EYA1	8	Branchio-oto-renal syndrome
2X FGFR2	10	Apert syndrome
FASL	10	Autoimmune lymphoproliferative syndrome
C1NH	11	Complement deficiency
AIP	11	AI porphyria
BRCA2	13	Breast cancer
NF1	17	Neurofibromatosis

2X and 3X refer to genes that have had two or three insertions, respectively.

of the role of the *Alu* element. Thus, at this point we believe that we underestimate the role of mobile elements in *recombining* and *damaging* the human *genome*.

AFTER EFFECTS

Alu elements may continue to contribute to genetic *instability* even if they do not initially damage a gene. A common form of secondary damage is owing to unequal *homologous* recombination, but several other types of mutations can alter the properties of *Alu* elements to create damage.

Alu elements may contribute to recombination events in several ways. The best understood is through either a deletion using the single-strand *annealing* reaction or a reciprocal, unequal homologous recombination, which can cause either duplications or deletions of the segments between the *Alu* elements that *recombine*. The recombination event does not have to be *reciprocal*.

For instance, if cells use the single-strand *annealing* pathway of recombination, nearby homologous Alu elements may recombine causing only deletions. These types of homologous recombination events have been estimated to cause at least 0.3% of human genetic disease.

However, as the *majority* of larger genomic rearrangements have not been characterized to this level, it seems likely that this represents an *underestimate*.

There are several studies suggestin *that Alu* elements located near one another in an inverted orientation may be even more destabilizing than those in a direct orientation. However, the *instability* caused by the inverted *Alu* elements does not create a predictable junction that allows individual recombination events to be definitively defined as being caused by *Alu* elements. Recombination can occur at various *locations* around both *Alu* elements.

Thus, it is difficult to distinguish between Alu-induced recombination events of this type, and Alu-unrelated *nonhomologous* end-joining events. However, it has been suggested that secondary structures may contribute to the majority of recombination junctions, and *Alu* elements represent one of the most *abundant* elements that could consistently contribute to such secondary structures.

We have previously reported a broad range of genes that have undergone *homologous* recombination events leading to genetic defects. These represent a broad variety of genes, as *Alu* elements are spread throughout *essentially* all genes. However, there are a few genes with

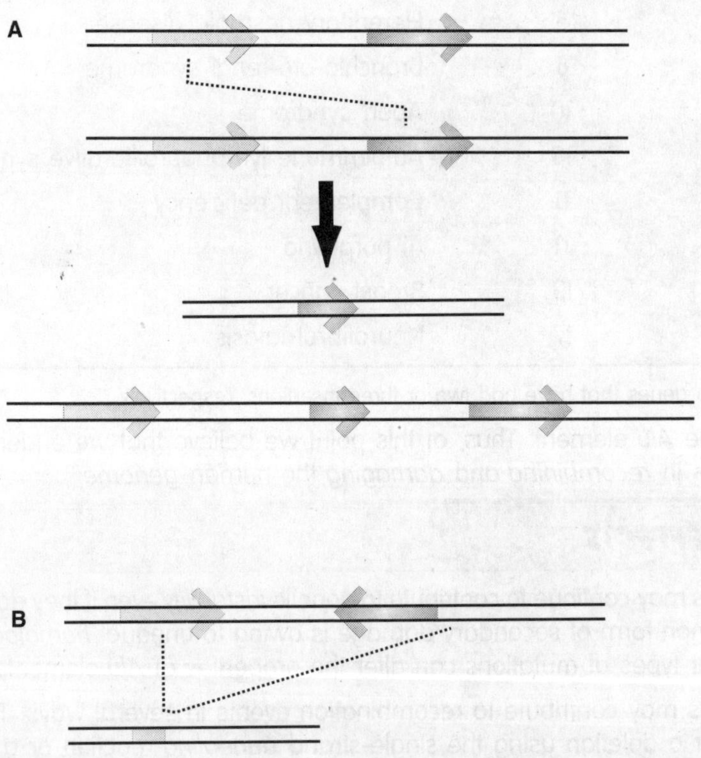

Figure 11.4: Variable effects: (A) When Alu elements are located in the same orientation near one another in the genome, it is possible for an unequal homologous recombination event to occur that can give rise to either a duplication or deletion of the sequences between the Alus. A hybrid Alu element is usually formed at the point of recombination. (B) WhenAlu elements are in the inverted orientation, they also seem to trigger recombination in their vicinity, but the recombination junctions do not seem to he driven by recombination so that the resulting products can be variable.

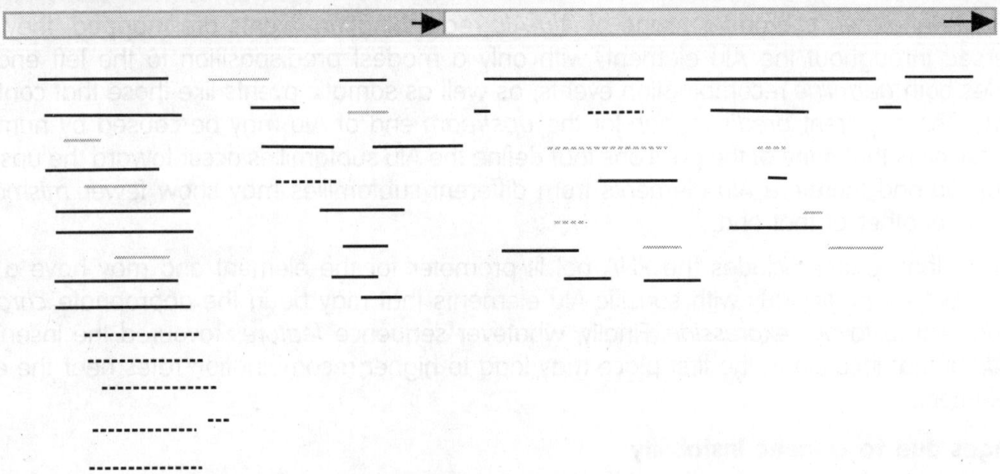

Figure 11.5: Graphic presentation of theAlu-Alu recombination junctions found in an assortment of recombination events. The darker dashed lines represent those occurring in theLDLR gene. The lighter dashed lines represent those occurring in the globin genes. The solid gray lines represent those that occurred in the MLL gene leading to acute myelogenous leukemia and the solid dark lines represent recombination events in an assortment of other genes. The length of the lines represents the uncertainty in the exact point of the recombination events relative to the schematic map of theAlu dimer shown. The arrows in the dimer portions represent the A-rich regions.

unusually high levels of *Alu Alu* recombination events. This includes a large number of events in the LDLR and C1 inhibitor loci.

However, there are also seven different recombination events leading to breast cancer in the *BRCA1* gene, three mutations in the *MSH2* gene that may represent as much as 10% of the defects in that gene, and a *duplication* and a deletion in *MSH6*. The *majority* of cases of acute *myelogenous leukemia* that do not involve a visible translocation have been shown to involve *Alu-Alu-mediated* duplication events in the *MLL* gene.

Although inter-chromosomal translocations generally do not involve *Alu*-Alu-*mediated* translocations, there is now some evidence that many of these cases have additional, smaller rearrangements in the *MLL* gene that may be Alu-mediated. Recently, it has been suggested that up to 30% of mutations in the Fanconi anemia gene *(FANCA)* may be caused by *Alu-Alu* recombination events and that these types of events are not typically detected using the polymerase chain reaction strategies commonly in use.

Thus, there is likely a strong ascertainment bias against detecting Alu-mediated genomic rearrangements. Most of *the Alu-Alu* recombination events are between nearby elements, generally spanning distances of less than 50 kb and often only a few kb in length. There are also a number of cases of chromosomal translocations that suggest some involvement of *Alu* elements, but very few that involve *Alu-Alu* homologous recombination.

It has been proposed that Alu elements may contain specific sequences that trigger recombination. An analogy has been made to *"chi-like"* sites that may cause targeting of the recombination events within the *upstream* portion of the *Alu* element. The original data that triggered that *hypothesis* were mostly from *recombination* events in the *LDLR* gene that appear to cluster.

However, when a broader range of *Alu-Alu* recombination events are mapped, they occur dispersed throughout the Alu element, with only a modest predisposition to the left end. This includes both *germline* recombination events, as well as somatic events like those that contribute to AML. The apparent *predisposition* for the *upstream* end of Alu may be caused by numerous factors. One is that none of the positions that define the Alu subfamilies occur toward the upstream end of *Alu* and therefore Alu elements from different subfamilies may show fewer *mismatches* with one another at that end.

Also, that region includes the RNA pol III promoter for the element and may have a more open *structure*, particularly with specific Alu elements that may be in the appropriate *chromatin* environment to favour *expression*. Finally, whatever sequence *features* favoured the insertion of the *Alu* at that location in the first place may lead to higher recombination rates near the end of the element.

Changes due to Genetic Instability

In addition to a major contribution to genetic instability through insertion and recombination, *Alu* elements may also contribute to other forms of DNA damage. There are a growing number of reports that *Alu* elements may influence splicing of genes. *Alu* elements have inserted in a number of loci without causing any apparent defect, but point mutations in the *Alu* elements have led to activation of *cryptic splice* sites causing genetic defects.

In addition, there is growing evidence that even *Alu* elements that are tolerated in genes may be *contributing* to a significant level of alternative and *aberrant* splicing. This leads to the inclusion of parts of *Alu* elements in a proportion of *transcripts* coming from the gene. In most cases these would be *expected* to result in defective RNAs and would, therefore, decrease the overall expression from those genes.

Although the A-tails of new *Alu* insertions appear to generally be homogeneous, they have been shown to commonly lead to the creation of more *complex microsatellites* over time. Although such changes would lead to frequent genetic *polymorphism*, the vast majority would be expected to be relatively *harmless*. One important example of a disease-causing change occurred in the middle A-rich region of an *Alu* in the frataxin gene.

In one particular human, this region mutated into a GAA *microsatellite*. This GAA microsatellite created the *permutation allele* that grew through triplet repeat instability to a size where it somehow blocked transcription of the frataxin gene, leading to *Friedreich ataxia* in those carrying this microsatellite. Other changes within these A-rich regions commonly generate the potential *polyadenylation* signal, AATAAA. This could lead to *truncation* of *transcripts* and *disruption* of gene expression.

Phylogenetic Effects of *Alu*

In order to create the more than 1 million *Alu* elements fixed today in the human genome, there have had to be massive numbers of other insertions events that were not fixed. Thus, *Alu* elements have been a major *contributor* to genomic instability and evolution throughout primate history. One _such event was an insertion-mediated deletion that inactivated the CMP-Nacetylneuraminic acid *hydroxylase gene* only in humans.

This led to altered protein *glycosylation*, which may have been a significant change helping to result in the *speciation* of humans from chimpanzees. *Alu-Alu* recombination events have also played an important role in chromosomal evolution and potentially speciation. It appears that an *Alu-Alu-mediated* recombination in the gulonolactone *oxidase gene* occurred after the divergence off *prosimians* from the other primates *t,05.*

This enzyme is a *critical* late step in the synthesis of vitamin C and resulted in the *inability* of primates to synthesize this vitamin, leading to the possibility of *scurvy.* In a broader sense, *Alu* elements appear to have been involved in a number of recombination events that have helped lead to segmental duplications on human *chromosomes.*

These segmental *duplications* have in turn been associated with a number of different syndromes through instability of the segmental duplications. Thus, *Alu* elements appear to have contributed to an overall rearrangement of the genome and *chromosomes* that has given rise to extra copies of genes, which may be beneficial for evolution, which on the other hand, can have negative impacts on the long-term *genomic stability* and function.

CONTROL

In addition to the evolutionary changes that *Alu* may cause that are related to its role in genetic instability, Alu elements have also been suggested to cause changes in gene structure and regulation. *Transcriptional* regulatory elements have been mapped to *Alu* elements near the promoters of genes, as well as *Alu* elements having been shown in several *reporter* systems to be able to contribute transcription factor binding sites to *stimulate* gene expression, as well as insulator sequences to *isolate genes* from other nearby elements.

Thus, *Alu* elements have probably influenced *expression* of many genes through insertion near their *promoters.* It also has been suggested that expression of *Alu* elements may contribute to a selective regulation of the initiation of translation. Because expression of *Alu* elements is stimulated by *viral infection, transformation, chemotherapeutic* DNA-damaging agents, and a number of *cellular stresses,* leading to speculation that this may help regulate the *translation* process in those situations.

Alu elements may make up a large portion of the *intronic* sequence in RNAs, as well as being presented in 3' non-coding regions. It has recently been noted that there are high levels of adenine-to-inosine RNA editing in *human cells,* and that more than 90% of it occurs within Alu elements. This is likely because of the ability of *Alu* elements in various *orientations* in the RNA to form *duplex* structures that make excellent substrates for the editing enzyme (ADAR).

Whether there is a role for this editing of *Alu* elements, or whether they just compete with other RNA substrates, is currently notknown. Through RNA editing, differential *splicing,* and other mechanisms, *Alu* elements may influence the processing and stability *numerous* cellular transcripts.

AMPLIFICATION

Although our understanding of both the mechanisms of *Alu* element *amplification* and their role in human disease have increased greatly, there are a number of important issues still to be resolved. In terms of human disease, one of the current problems is that many of the polyrnerase chain reaction-based strategies used to *identify* human disease mutations are biased against the detection of large sequence *insertions* or *deletions*. Thus, we are probably not detecting a significant proportion of the *Alu* element-induced damage in human disease.

As we learn more about the mechanism of Alu amplification, it will be important to focus on which cell types are involved in the *amplification*; *germ line, somatic*, and *tumor cells*. Also important are questions of *howAlu* elements interact with cellular gene products that may modulate, or respond to, their amplification.

This will help determine whether different individuals are more or less susceptible to damage by these elements. Similarly, with evidence that expression of *Alu* elements being stimulated by *numerous* environmental factors, the possibility that different individuals are more or less prone to damage by Alu elements in response to genomic exposures will be an important issue. Finally, Alu elements apparently utilize the L1 *machinery* for their amplification more efficiently than does L1 itself.

It will be important to delineate both the *commonalities* and differences in the actual mechanism of amplification of these *elements*, as well as their respective impacts once inserted in the genome, in order to fully assess the roles these elements play in human genetic instability. Ultimately, we must determine whether there is anything we can do to *minimize* the genetic *instability* caused by these elements.

Alu elements have caused *insertional* mutation events throughout primate evolution and continue to cause a significant level of sporadic human genetic disease. Furthermore, the high copy number of *Alan* elements in the human genome contributes to *secondary* events, such as *recombination* events, that also contribute extensively to human disease. There are also a number of other ways in which *Alu* elements contribute to human genetic diversity (i.e., causing alternative splicing and changes in gene regulation).

Chapter 12

Process of Identification

The identification of single *nucleotide* changes in DNA sequences is critical to many fields of *biological* research, from areas as diverse as the *characterization* of single nucleotide *polymorphisms* (SNPs) to identifying specific *alterations* caused by random *mutagenesis* screens. There are many approaches which can be used to compare DNA sequences for such *nucleotide* substitutions, ranging from the most basic, of comparing plain text sequence to a reference, to complex *multifunctional analysis* software.

However, these can often suffer from a *compromise* between *simplicity* and *efficiency*, where the simpler methods may generate false-positive and false-negative results, whereas the more accurate methods can be *overly complex* and *time-consuming* for small-scale analysis. SeqDoC is a web-based application, which provides a way of identifying *nucleotide substitutions* with high accuracy and good *sensitivity* through a very simple and *intuitive* interface.

It does this by generating a *"difference profile"* between pairs of sequence *chromatogram* traces, readily *highlighting* any changes characteristic of single *nucleotide* substitutions. The user then has only to examine the sequence traces at these specific points to determine the nature of the *substitution*; this *intervention* avoids the incorrect calls, which can result from fully automated software.

Simplicity of use is enhanced the use of all *predefined* parameters, which have been optimized to produce the highest signal to noise ratio for features indicative of *nucleotide* substitutions. SeqDoC also *compensates* for sequencing artefacts including sequence *compression*, peak height *variation*, drop off of signal strength, and offset start positions of the sequences.

REQUIREMENTS

Feeding

Input for SeqDoC is two (or optionally more) Applied *Biosystems* (ABI) format sequence *chromatogram* files. Many other sequence formats (including SCF and MegaBACE files) can be converted to ABI format using free conversion software such as Chromas Lite (http://

www.technelysium.com.au/ chromas lite.html). One file should be a known reference sequence, the other(s) are test sequences to be compared against the reference. SeqDoC directly compares the chromatogram traces themselves between the sequences being examined, so no manual correction of base *calling* is required prior to *testing.*

Web Based Tool

SeqDoC is primarily a web-based tool, and can be found at http:// research. imb.uq.edu.au/ segdoc/. It can be *accessed* using any Internet-enabled *computer,* and works with all major *operating systems* (Windows, MacOS, and Linux) and web-browsers (including Firefox, Internet Explorer, Mozilla, and Netscape). There are no restrictions on the use of the web-based version.

Sources Used

The source code for SeqDoC can be downloaded from the SeqDoC website (http:// research.imb.uq.edu.au/seqdoc/) or *Bioinformics.*Org (http:1/ bioinformatics.org/project/?group id=481) for installation on a local server.

In this case it will be necessary to install the GD graphics library (http:// www.boutell.com/gd/) and the Perl GD, GD::Graph and ABI modules (http:// search.cpan.org/dist/GD/, http:// search.cpan.org/dist/GDGraph/, http://search. cpan.org/dist/ABI/). SeqDoC is made available under the terms of the GNU General Public License (http://www.gnu.org/licenses/gpl.html).

PROCEDURES

There are two versions of SeqDoC available at the website. The standard *version* should be used for direct comparison of two sequences, whereas multiple SeqDoC allows the user to *simultaneously* compare up to ten *sequence* traces against a single reference trace. The two versions use *identical* processing algorithms, and *interpretation* of output is the same for both.

The interface differs only by the number of sequence *chromatograms* which can be uploaded. Multiple SeqDoC has two primary benefits for the user when three or more sequences are to be compared: first, all sequences are aligned to the same reference and are displayed on the same output page, simplifying *interpretation* of results; second, there can be a performance benefit of up to 30% decrease in processing time per sample.

Programming

SeqDoC has no user-definable parameters, the only necessary steps being to upload a reference sequence file and one or more test sequences files, depending on which version of SeqDoC is being used. The file *upload* is carried out using the upload boxes on the webpage, and the user then simply clicks the "*submit query*" button to perform the comparison. Results will typically be returned after around 15 s for a single *comparison,* and *proportionally* longer for multiple comparisons.

The reference and test sequences should of course be of the same region of DNA, and should have been sequenced in the same *orientation*; relative peak heights will be *inconsistent* in samples sequenced in opposite directions because of *variations* in sequencing chemistry and are likely to prevent usable results being returned. It is preferable, although not necessarily essential, that the sequences being compared have been sequenced using the same primer.

Results

The output provided by SeqDoC is three aligned images in a web browser window or, for multiple SeqDoC, three aligned images per test sequence. The top image of each three is the reference *sequence* trace and the *bottom image* is the test sequence. Both are displayed as png images and cannot be reconverted back to chromatograms. The "*difference profile*" is provided as an aligned png image between the two sequence traces.

All three traces are colored according to standard ABI chromatogram colors (A - green, G - black, C - blue, T - red). In the difference trace, values above zero indicates the channel which is stronger in the reference sample, while values below zero means that the signal for that channel is stronger in the test sequence. The actual Y-coordinate values in all three traces are arbitrary values resulting from the *normalization* routine and difference *calculation algorithm*.

Base position counts are included in both sequence trace images. These correspond directly to positions on the input sequence and, because the start of the test sequence may have been shifted to align the two traces, these position counts may not match directly between the aligned traces.

The output traces are subjected to *normalization* and alignment *adjustment*, and so will *typically* appear subtly different to the input traces; however, this will only affect peak spacing and height, and will not *fundamentally* alter the trace.

Discussion

Substitutions and Indels of Nucleotide

SeqDoC does not provide any automatic classification of *nucleotide* substitutions beyond the algorithm for emphasising base substitutions over sequence noise; instead interpretation of the difference trace is left to the *researcher*. This *interpretation* is largely intuitive-the difference trace overall should be *largely flat*, with any major deviations *indicative* of a difference between the *reference* and *test traces*.

Because the two sequence traces are aligned with each other and with the difference trace, the investigator need only look at the sequence peaks immediately above and below the point of difference to establish the nature of the difference. The most frequently observed features are single *nucleotide substitutions*. These result in large *bidirectional* peaks in the difference profile as seen in Figure elsewhere in this chapter and, in more detail, in Figure elsewhere in this chapter.

Examination of the reference and test traces at the point of this bidirectional peak will allow the *researcher* to identify the substitution as well as its coordinate positions within the original sequence traces. When a *mixed peak* occurs in the test sequence, such as that caused by a *heterozygous* SNP, a *bidirectional* peak will still be observed in the difference trace.

In addition to highlighting base substitutions, SeqDoC can also be used to rapidly find single base *insertions* or *deletions* (*indels*). These are characterized by a sudden occurrence of major differences between the traces from the point of the indel (when *nonmatching* bases start to be compared) followed by a return to minimal difference after 8-10 bases once the alignment has been restored by the software.

Characterization of indels is similar to that for nucleotide substitutions; the *investigator* simply needs to examine the two sequence *traces* at the start of the region of *misalignment* to identify the inserted or deleted base.

Diagnosing Messages

The SeqDoC interface performs basic error checking on input sequences, and provides

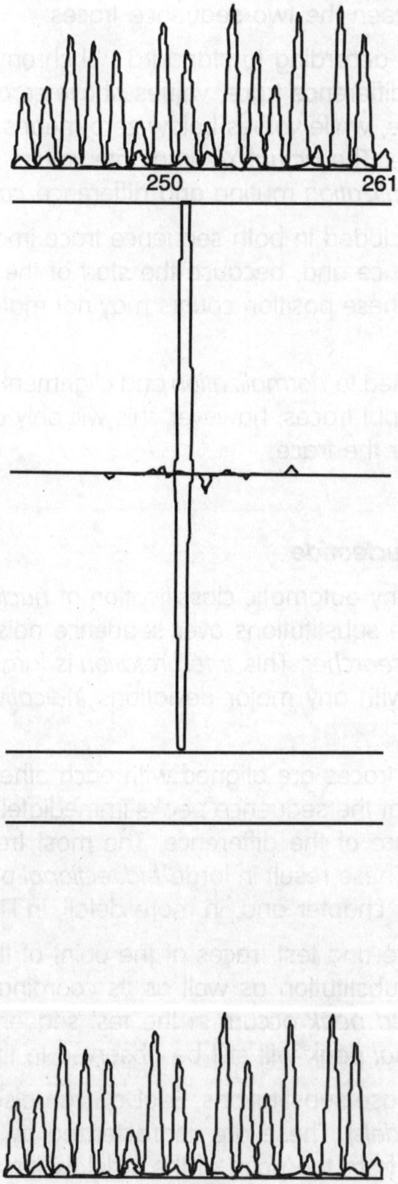

Figure 12.1: The characteristic bidirectional peak caused by a single nucleotide substi-tution (in this case a G at position 251 in the reference sequence has changed to an A at position 249 in the test sequence).

diagnostic messages where possible (for example, when a file does not appear to be in ABI format). However, no checking is carried out on the output and *familiarity* with some of the problems that can occur can save much time and effort when an *unexpected* result is obtained.

Probably the most frequent cause of failure is uploading two sequences that cannot be aligned.

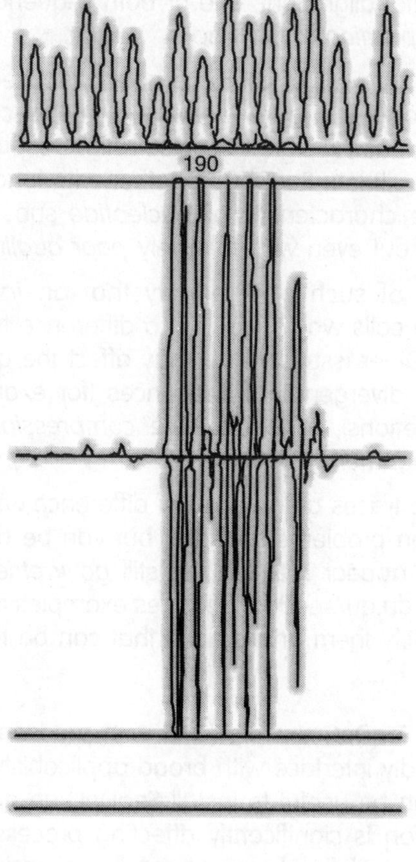

Figure 12.2: The pattern resulting from a single base insertion/deletion. Here, a G at position 189 in the reference sequence is missing from the test sequence, causing the traces to shift temporarily out of alignment.

Usually this would occur because an incorrect sequence has been selected by accident, but it can also result from the start points of the two sequences being too far apart for the initial *alignment* to succeed.

In such cases, rather then being flat, the difference profile will be very noisy for the whole length of the traces. If an incorrect sequence has been selected, this is easily resolved, but if it is caused by *problems* with the initial alignment, one or both sequences will need to be *manually* edited to better match the start *position* of the traces.

Poor quality input sequence data will of course produce reduced quality of output data with a noisy difference trace, as shown in Figure elsewhere in this chapter, and will complicate interpretation because of the decrease in signal to noise ratio. Despite this, the *normalization* and corrective alignment *algorithms* built into SeqDoC, together with functions to enhance features of the difference profile which are characteristic of *nucleotide* substitutions, means that useful comparisons can still be carried out even with relatively *poor quality* sequence.

Typically, only a sequence of such poor quality that an *investigator* could no longer *unambiguously* identify the base calls would result in a difference trace that was so noisy as to obscure *genuine* signal peaks. Other issues which may affect the quality of data obtained from SeqDoC include dye-blob noise, divergence of sequences (for example by *co-ligation*, or as a result of large insertions or deletions), and sequence *compression* that is too extreme to be compensated for by the *auto-alignment* function.

Examination of the sequence traces at the point of difference will readily identify the first two situations. Sequence compression problems are rare, but can be difficult to diagnose, because the two sequences are likely to appear identical yet still go *irretrievably* out of alignment. Our website (http://research.imb.uq.edu.au/seqdoc/) features examples of a number of such problems with information on how to identify them and actions that can be taken to rectify them.

Applicability of Installation

The web-based versions of SeqDoC are provided with predefined settings so as to provide the simplest and most user-friendly interface with broad applicability. However, there are several circumstances under which it can be useful to install SeqDoC on a local computer, for example where a slow network connection is significantly affecting processing time, where proprietary data is being used, or where the predefined settings are *inappropriate* for a specific *analysis* and need to be modified.

Settings which can be altered include increasing the window size for the initial alignment (to simplify *comparison* of samples sequenced with different primers), modifying the auto-alignment function to compensate for sequence compression, changing the difference trace calculations to increase the sensitivity of mixed peak detection, and even changing output trace colors to assist *researchers* with color blindness.

SeqDoC is a Perl program (http://www.perl.org/) and should work on any computer capable of running Perl. A number of accessory programs are also required to run SeqDoC-these are the GD graphics library and the Perl GD, GD::Graph and ABI modules. In addition, web server software is required to use the web interface (the public SeqDoC version uses the Apache HTTP Server), although the code can be modified to generate a stand-alone application.

Full instruction as to installation of these programs is outside the scope of this article, but interested researchers are welcome to contact the author for advice. This initial setup of SeqDoC requires a reasonable *proficiency* with building *libraries* and software from source, but subsequent *modifications* of parameters can be performed by anyone with basic Perl programming skills.

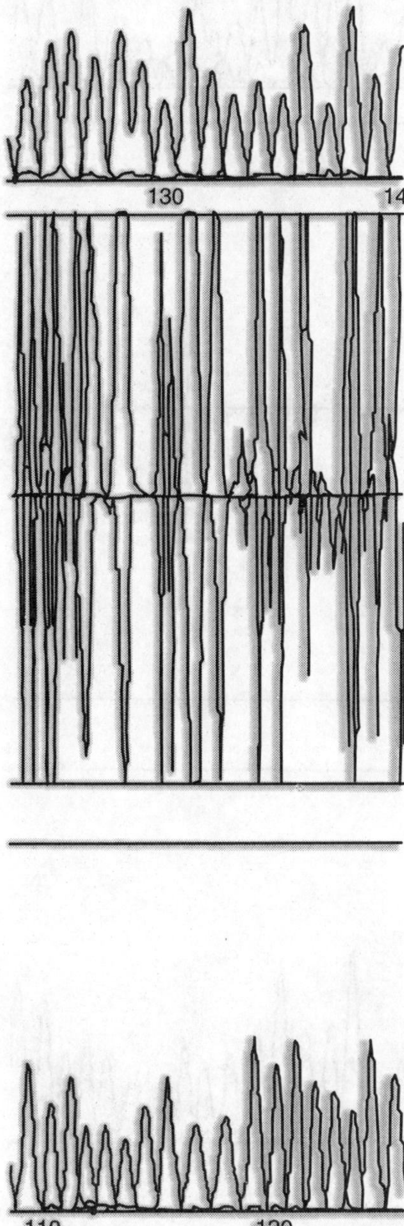

Figure 12.3: Nonmatching sequences cannot be aligned, so give a very noisy difference trace.

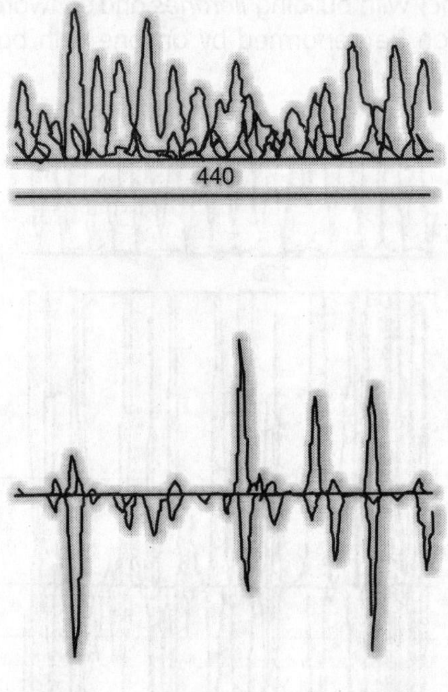

Figure 12.4: Poor quality sequence data (top trace) results in a noisy difference trace.

POINTS TO REMEMBER

1. The increase in speed when using multiple SeqDoC is proportional to the number of test sequences. A two sequence comparison (i.e., with one reference and one test sequence) will take the same time using either version of SeqDoC, whereas the maximum 30% increase will be obtained when using 10 test samples.
2. For a comparison of two sequences, either can be assigned as the reference; the differences will be identified equally well. Processing of the two sequences is identical except that *alignment* alterations (insertion/deletion of sequence at the start of the trace for initial alignment and subsequent adjustments to peak spacing) are made only to the test sample.
3. For a multiple sequence comparison, the reference should be the sequence file from the wild-type/default sample if there is one, otherwise a good quality sequence trace with no mixed base calls (such as heterozygous SNPs). Although mixed bases in the reference do not have any specific impact on *generation* of the difference trace, they may complicate the subsequent analysis of the output.
4. The processing time will be significantly increased if a slow network connection is being used, because of the length of time taken to upload the files to the server.
5. Auto-start alignment can compensate for misalignments of up to around 20 nucleotides between the start positions of the two sequences, so nearly adjacent primers (such as in *cloning vectors*) should also be acceptable. *Alternatively,* initial manual alignment can be performed with chromatogram processing software, *providing* that the modified sequence traces are saved as ABI format files.
6. *Bidirectional* peaks in the difference trace are indicative of base substitutions because the value for one channel (that corresponding to the base in the reference sequence) will be very much higher in the reference than the test at that point in the trace, whereas conversely the value for the channel corresponding to the base in the test sequence will be very much lower.
7. A *bidirectional* peak will be observed in the presence of a mixed base because, even though the base from the reference sequence is still present, it will be of reduced relative *intensity* in the test sequence.

   The magnitude of the difference trace peak for a 50/50 mixed base is *theoretically* approx 20% of that of a complete substitution. However this is subject to many other factors, and the magnitude of the difference peak is not a *reliable* method of *determining* the nature of the *substitution.*
8. If testing pooled DNA for SNPs, allele frequencies of around 20% or more can typically be detected. The default parameter settings of SeqDoC are optimized for substitutions, and *consequently* rarer alleles are not distinguished from noise (Dr. J. Logan, personal communication). The *sensitivity* of SeqDoC for such applications can be altered by installing a local version of the *software* and *modifying* the parameters to find the most suitable *optimization* settings.

9. Noise peaks do tend to appear different to the *bidirectional* peaks characteristic of nucleotide substitutions, in that they are often unidirectional or at least asymmetric, but they can also appear as bidirectional peaks very similar to genuine substitutions, and each of these will need individual examination.

10. Variation of gel conditions and DNA templates can, on occasion, cause the peak spacing interval to vary too greatly between the reference and test samples for the alignment algorithm to keep the two traces aligned. This may sometimes be corrected by modifying sequencing conditions, or can be compensated for by installing SeqDoC locally and modifying the alignment settings appropriately (Dr. A. Topf, personal communication).

11. Problems with sequence compression are probably best diagnosed by a preliminary comparison of the two sequences with a standard chromatogram viewer set to the same horizontal scale for both traces. A difference in base count of more than around 15% over the same size window may indicate sequence compression.

Chapter 13

ncRNA and RNAz

In contrast to protein-gene finders that are *routinely* used for genome annotation, noncoding RNA (ncRNA) gene finders are still in their infancy. *Systematic de novo* prediction of ncRNAs is hindered by the fact that there are no common *statistically* significant features in primary sequence (e.g., open reading frames or codon bias), which could be *exploited* for efficient *algorithms*. It is not clear what is defined as "ncRNA."

There is no doubt that independent "*RNA genes*" with a defined *molecular* function, such as tRNAs, microRNAs, or snoRNAs, should be called ncRNAs. But the situation is not always clear. The transcriptional activity of *mammalian genomes* is much more complex than anticipated. We see mRNA-like ncRNAs, non-polyadenylated RNAs from both intronic and intergenic regions, overlapping transcripts, extensive antisense *transcription*, and *transcribed protein-pseudogenes*.

In addition, there is a recent example of a noncoding transcript that only is expressed to interfere with and *downregulate* the transcription of a *neighboring* gene, but the produced RNA molecule itself does not have any obvious function. There is even an example of a functional RNA encoding a protein. The spectrum of ncRNAs and their mode of action is very *heterogeneous*. It can be safely assumed that the full spectrum of functions is not yet discovered and that a general ncRNA gene finder is an *unrealistic* goal, even in the long term.

However, there is a subclass of ncRNAs that-with the help of *comparative genomics* can be predicted with fair *accuracy*. Structural ncRNAs have a defined and *evolutionarily* conserved secondary structure that is of functional importance. Most of the well-known "classical" ncRNAs, as for example tRNA, rRNA, RNAse P, or SRP RNA, are of this class. *Pioneering* work in the prediction of structural ncRNAs by *comparative genomics* was preformed by Rivas and Eddy.

QRNA predicts conserved RNA secondary structures on *pairwise alignments* using a probabilistic approach based on a *stochastic* context free grammar to model RNA structure. RNAz takes a different approach. It is based on minimum free energy (MFE) structure prediction algorithms. It relies on the fact that structural RNAs have two charac-teristic features: (1) unusual *thermodynamic stability* and (2) *conservation* of secondary structure. The following section outlines the basic principles of RNAz.

RNAz

It is easy to calculate the MFE as a measure of thermodynamic stability for a sequence using, e.g., RNAfold. However, the MFE depends on the length and the base composition of the sequence and is, therefore, difficult to interpret in absolute terms. RNAz calculates a normalized measure of thermodynamic stability by *comparing* the MFE m of a given (native) sequence to the MFEs of a large number of random sequences of the same length and base *composition*.

A *z*-score is calculated as $z = (m - \mu)/\sigma$, where μ and σ are the mean and standard deviations, respectively, of the MFEs of the random samples. Negative *z*-scores indicate that a sequence is more stable than expected by chance. RNAz does not actually sample random sequences but approximates *z*-scores, which is faster but of the same accuracy.

RNA Alifold

RNAz predicts a consensus secondary structure for an alignment by using the RNAalifold approach (RNAalifold works almost exactly like single sequence folding algorithms (e.g., RNAfold), with the main difference being the energy model is augmented by covariance information. *Compensatory* mutations (e.g., a CG pair mutates to a UA pair) and consistent mutations (e.g., AU mutates to GU) give a *"bonus"* energy, whereas *inconsistent* mutations (e.g., CG mutates to CA) yield a penalty. This results in a *consensus* MFE E_A.

RNAz compares this consensus MFE to the average MFE of the individual sequences $\bar{E}$ and calculates a structure conservation index: $SCI = E_A/\bar{E}$. The SCI will be high if the sequences fold together equally well as if folded individually. On the other hand, SCI will be low if no consensus fold can be found.

Diagnostic Features

The two independent diagnostic features of structural ncRNAs, *z*-score and SCI, are finally used to classify an alignment as *"structural RNA"* or *"other."* For this purpose, RNAz uses a support vector machine (SVM) *learning algorithm*, which is *trained* on a large test set of well known ncRNAs.

Using RNAz, it is possible to efficiently screen alignments for functional RNA *secondary structures*. It is important to note that RNAz cannot distinguish functional RNA elements that are part of ncRNAs from elements that are cis-regulatory elements of mRNAs.

MAIN GOALS

There are two main goals of this chapter. First, we want to give detailed technical advice on how to use RNAz. Second, we want to provide the user with a well-founded understanding of the results from RNAz. We want to assist in a sensible *interpretation* of RNAz *predictions-leading,* as we hope, to *reasonable conclusions* for the application.

In the first part, we explain how to install RNAz and all necessary helper programs on the system. Next, we demonstrate the basic usage of RNAz including the correct formatting of the input alignments. More advanced techniques, which require *preprocessing* steps of the input alignments, are discussed afterward. In the last section, we demonstrate how to conduct a RNAz screen of a large number of *automatically* generated alignments on the example of genome-wide screen of *Saccharomyces cerevisiae.*

Topography

There is no graphical user interface for RNAz. All steps are carried out in a command-line (terminal). Lines starting with a "#" are commands and they should be typed into the terminal window, followed by *pressing* return. The "#" sign stands for the command-line prompt and may not look of same on different systems. If a command is too long for one line in this chapter, it is separated by a backslash "\" and continues on the next line.

Do not input the *backslash*, simply type in the command on one line. All programs are implemented as filters, i.e., they read from the standard input and write to the standard output. Therefore, we make use of the pipe and *redirection* operators ("<", ">").

Online documentation on the usage of each program can be obtained by using the — help option, e.g.:

```
# RNAz-  -help
```

Most command-line options have a long (e.g., —help) and a short (e.g., -h) form. For *didactic reasons,* we use long option names throughout this chapter.

REQUIREMENTS

1. Hardware. RNAz is generally fast. Small- to medium-sized data sets can be analyzed within reasonable time on a single modern desktop or *laptop computer.*
2. Operating system. If available, we recommend to use a Linux/UNIX system for the analysis. Also Mac OS X, in principle a full-featured UNIX system, is an *adaequate platform.* Alternatively, RNAz can also be run on Microsoft Windows. Most of the methods described in this chapter can be carried out on Windows without any modification.
3. Perl. The RNAz program is bundled with a variety of helper programs, which are written in the Perl programming language. To run these programs, Perl needs to be installed on the system, which is most likely the case in all Linux/UNIX systems and Mac OS X. Perl is not part of a standard Windows system. Windows users can download it from www.activestate.com.

   Choose the latest ActivePerl MSI installer package for Windows and simply follow the *installation* instructions. Make sure that selected the "Add Perl to the PATH environment variable" and "Create Perl file extension association" options have been selected during installation.
4. RNAz. The RNAz program can be downloaded from www.tbi.univie.ac.at/^wash/ RNAz. For the examples in this chapter, RNAz v 1.0 was used. For Linux/UNIX and OS X, download the file *RNAz-1.0 . tar. gz.* Windows users can download the file *RNAz-1 .0-win32.msi.*
5. Optional software. Some advanced analysis steps require additional software to be installed on the system. To create HTML formatted output of the results as described in elsewhere in this chapter the Vienna RNA package (www.tbi.univie.ac.at/RNA) and the postscript

interpreter Ghostscript (http://www.cs.wisc.edu/^ghost/) need to be installed. To perform automatic database searches of predicted ncRNA candidates the user will need NCBI Blast (ftp://ftp.ncbi.nih.gov(blast).

6. Example files. Most of the example files used in this chapter are part of the RNAz package. To reproduce the *S. cerevisiae* screen described in elsewhere in this chaper the data file can be downloaded file from www.tbi.univie.ac.at/ papers/SUPPLEMENTS/MiMB/.

PROCEDURES

In the simplest case, run the following series of commands to build and install RNAz:

```
# tar -xzf RNAz-1.0.tar.gz
# cd RNAz-1.0
#. /configure # make
# su
# make install
```

This requires root *privileges* and installs all files under the /usr/local tree. The RNAz executable is *installed* in /usr/local/bin and the user should now be able to run the program (try RNAz —version on a *terminal window*). If the user does not have root *privileges* or *experiences* other problems (e.g., gcc compiler not found).

The Perl programs are installed to /usr/local/share/RNAz/perl. To make these programs available from other locations, the user can either add this directory to the PATH of *executables* environment variable or copy the Perl programs to an existing directory already in the PATH.

Microsoft Windows

To install RNAz on Windows simply double click on the RNAz - 1 . 0 - win32.msi and follow the instructions. Open a command prompt and type `RNAz—version` to test the installation.

Loading Optional Software

We cannot cover in detail the installation procedure of the optional software. We will just give an outline of how to install the Vienna RNA package and NCBI blast on a standard *Linux system*.

To install the Vierina RNA package, get the latest `ViennaRNA-X.X.tar.gz` file from www.tbi.univie.ac.at/RNA. The package can be installed in exactly the same way as RNAz, using ./configure and make. Please refer to the INSTALL document for detailed installation options. Make sure that the Perl programs in the Utils directory are in the PATH of *executables*.

To install NCBI Blast, download the `blast-2.*.tar.gz`-package matching the platform from ftp://ftp.ncbi.nih.gov/blast/executables/LATEST/. Copy it to the installation directory of choice and "*untar*" it. The executables are located in the bin subdirector, which should be added to the PATH variable.

Loading Required Files

Move the example file yeast-examples. tar.gz to the directory of choice and "untar" the file:

```
# tar -xzf yeast-examples.tar.gz
```

A

```
CLUSTAL W (1 .83) multiple sequence alignment


sacCer1    GCCTTGTTGGCGCAATCGGTAGCGCGTATGACTCTTAATCATAAGGTTAGGGGTTCGAG
sacBay     GCCTTGTTGGCGCAATCGGTAGCGCGTATGACTCTTAATCATAAGGTTAGGGGTTCGAG
sacKlu     GCCTTGTTGGCGCAATCGGTAGCGCGTATGACTCTTAATCATAAGGTTAGGGGTTCGAG
sacCas    GCTTCAGTAGCTCAGTCGGAAGAGCGTCAGTCTCATAATCTGAAGGTCGAGAGTTCGAA
               ** *    * ** ** **** ** ****   * *** *****   ****     * ****** *

sacCer1       CCCCTACAGGCT
sacBay        CCCCTACAGGCT
sacKlu        CCCCTACAGGCT
sacCas        CTCCCCTGGAGCA
```

B

```
##maf version =1
a  score=119673 . 000000
s  sacCer1 . chr4        13522453  73    -   1531914 GCCTTGTTGGCGCAATCGGTAGCGCGTATGACTCTT ..
s  sacBay . conting_465     14962  73    -     57401 GCCTTGTTGGCGCAATCGGTAGCGCGTATGACTCTT ..
s  sacKlu .  Conting1694      137  73    +      4878 GCCTTGTTGGCGCAATCGGTAGCGCGTATGACTCTT ..
s  sacCas .  Conti      128   258  73  +         663  GCTTCAGTAGCTCAGTCGGAAGAGCGTCAGTCTCAT ..
```

C

```
########################    RNAz  0.1.1  ########################

Sequences:  4
Columns:  73
Reading direction: forward
Mean pairwise identity:  80-82
Mean single sequence MFE:  -27.20
Consensus MFE:  -26.50
Energy contribution:  -23.62
Covariance contribution:  -2.88
Combinations/Pair:  1.43
Mean z-score:  -2.18
Structure conservation index:  0.97
SVM decieion value:  2.39
SVM RNA-class probability:  0.993311
Predication:  RNA

#

>sacCer1 . chr4    1352453    73   -  15319114
GCCUUGUUGGCGCAAUCGGUAGCGCGUAUGACUCUUAAUCAUAAGGUUAGGGGUUCGAGCCCCCUACAGGGCU
(((((((.(((((........))))...((((.((((....)))))))) (( ((((....)))))).))))))) . (-29.20)
>sacBay  . conting _  4651496273  - 57401
GCCUUGUUGGCGCAAUCGGUAGCGCGUAUGACUCUUAAUCAUAAGGUUAGGGGUUCGAGCCCCCUACAGGGCU
(((((((.(((((........))))...((((.((((....)))))))) (( ((((....)))))).))))))) . (-29.20)
>sacKlu. Contig169413773 + 4878
GCCUUGUUGGCGCAAUCGGUAGCGCGUAUGACUCUUAAUCAUAAGGUUAGGGGUUCGAGCCCCCUACAGGGCU
(((((((.(((((........)))).((((.((((....))))))))(((((....)))))).))))))).(-27 .20)
>sacCas . Contig128 258 73 + 663
GCUUCAGUAGCUCAGUC
(((((((..((((........)))).((((.........))))((((((......)).))))...))))))). (-23020)
>consensus
GCCUUGUUGGCGCAAUCGGUAGCGCGUAUGACUCUUAAUCAUAAGGUUAGGGGUUCGAGCCCCCUACAGGGCU
((((( ((..((((........)))).(((((.. .....)))) ).....((((((.......))))))))) ))).
                                        (-26.50  =  -23.62  +    -2.88)
```

Figure.13.1: Supported alignment formats and RNAz output. (A) Clustal W format, (B) MAF format (sequences have been shortened because of space restrictions), (C) output of RNAz on the MAF file shown in B.

If Windows is being used, download the file yeast-examples.zip and unzip it in the directory of choice.

Basic Usage of RNAz

Preparation Alignment

RNAz takes a multiple sequence alignment as input. RNAz does not align sequences, so other programs need to be used to create the alignments. If the alignments are prepared manually (in contrast to *automatic* genome-wide alignments), we recommend using Clustal W. It is an easy-to-use and widely available tool, which performs well on structural RNAs.

For hints on preparing the alignments RNAz can read two different alignment formats: Clustal and MAF. The Clustal W format is a *concise* format, which is supported by many programs and thus suitable for everyday use.

For genomic screens, however, it is necessary to exactly store the genomic locations of aligned sequences. For this purpose, the MAF format was developed, which requires six fields for each sequence entry:

1. A unique identifier of the source sequence.
2. The start position of the aligned subsequence with respect to this source sequence.
3. The length of the aligned subsequence without gaps.
4. A "+" or "–" indicating if the sequence is in the same reading direction of the source sequence or the reverse complement.
5. The sequence length of the complete source sequence. 6. The aligned subsequence with gaps.

The full specification of the format can be found at http://genome.ucsc.edu/goldenPath/help/maf.html. Note that RNAz and all other helper programs do not make use of field 5 and also ignore the value of the "score=" field in the header line. So it is possible to simply fill these fields with 0 or any other arbitrary values, if the real values are not easily available.

The RNAz package contains several example files, which are by default installed to /usr/local/share/RNAz/examples. To run the following examples change into this directory.

Working with RNAz

As soon the alignments have been prepared, they can be immediately scored with RNAz. In the simplest case, type:

```
# RNAz tRNA.maf
```

The file tRNA. maf is that one shown in Fighure esewhere in this chapter and the command gives the output shown in Fighure elsewhere in this chapter.

RNAz Output

As described in the introduction, RNAz calculates various folding charac-teristics to classify the alignment. These are displayed in the header section of the RNAz output. The mean single MFE is compared to the *consensus* MFE, which results in the SCI, a measure for structural *conservation*

.In this ideal example of a tRNA, we observe a very high SCI of 0.97. The SCI depends on the mean *pairwise* identity and the number of sequences in the *alignment*.

So, it is not possible interpret the significance of a SCI-value in absolute terms. As a rule of thumb, a SCI near or even more than the mean pairwise identity is "*good*" and might indicate structural conservation. For example, given an alignment with five sequences and a mean *pairwise* identity of 60%, a SCI of 0.75 can be regarded as strong hint for a conserved fold. On the other hand, on a pairwise alignment with 90% identity, SCI = 0.75 does not indicate a conserved fold at all.

The second characteristic is thermodynamic stability, which is expressed as the mean *z*-score of the sequences in the alignment *z*-scores of MFEs are not normally *distributed*, so the user cannot directly give a statistical significance for the *z*-score. However, mean *z*-scores less than –3 or –4 generally indicate very stable structures that should arise only in rare cases by chance.

Also here, one has to consider the overall sequence divergence in the alignment. On a pairwise alignment with 90% identity a *z*-score of –4 is much more likely to occur by chance than on an alignment of six sequences with only 60% identity. Apart from SCI and *z*-score, there are a few other values displayed in the RNAz output (for their meaning 6).

RNAz assists in the final classification by providing an overall "RNA-class probability," or "*p*-value." It is important to know that this is not a *p*-value in a strict statistical sense, simply because there is no underlying statistical model. Instead, RNAz uses a rather *ad hoc* machine learning technique to calculate this value. If $p > 0.5$, the alignment is classified as "RNA." The false-positive rate at this cutoff was found to be ~4 %, i.e., we expect four positive hits in 100 random alignments.

For many applications it is useful to set a more stringent cutoff of p = 0.9 with an associated false-positive rate of ti 1 %. It turned out to be a useful practice to use $p = 0.5$ and $p = 0.9$ as two main levels of *significance*. A more sophisticated interpretation of the *p*-value without considering the other values is generally not useful. In most cases, the user cannot say that, for example, a hit with $p = 0.97$ is more reliable than a hit with $p = 0.95$.

In the lower part of the RNAz output the predicted structures are explicitly seen for the sequences. There will be structure predictions for each single sequence and a consensus structure prediction for the whole alignment. The *predicted* structures are given below the sequences in a "*dot-bracket*" notation.

Each basepair in the secondary structure is indicated by a pair of brackets: "(" and ")". Unpaired bases are shown as dot: ".". Next to the structure the MFE will be seen in kcal/M. A graphical output can be obtained by using *RNAalifold*, which is described in detail in Chapter elsewhere in this chapter of this book.

Modernization of RNAz

For a given alignment, a putative RNA can either be read in the forward direction or in the reverse complementary direction. Therefore, both reading directions should be scanned. By default, only the forward direction is scored, but the *—forward, —reverse,* and *—both-strands* flags can be used to explicitly specify the reading direction.

If there is a strong RNA signal in one strand, in many cases, a signal can also be observed in the reverse complement. Usually the signals (SCI, *z*-score, consensus MFE) are stronger in the "correct" direction. In most cases, this also goes along with a better *p*-value. That is not always the case and, therefore, RNAz uses a separate SVM decision model to predict the correct strand. Please note, that in v 1.0 this is still an *experimental* feature. The following command will *analyze* both strands of the tRNA and, in addition, activate the strand *prediction*:

```
# RNAz--both-strands--predict-strand tRNA.maf
```

In this example, the signal from both strands are almost indistinguishable and also the p-values are almost the same (0.993 and 0.999). RNAz still suggests the correct (*forward*) strand and displays a "strand class probability:"

```
# Strand winner:forward (0.88)
```

RNAz is currently limited to *alignments* with not more than six sequences. If there are more than six sequences in the alignment, the number must be reduced either manually or by using the rnazSelectSeqs. pl program to filter the alignment before it is put into RNAz:

```
# rnazSelectSegs.pl miRNA.maf|RNAz
```

The file miRNA. ma f contains 12 aligned microRNAs. With default param-eters, rnazSelectSeqs .pl selects a subset of six sequences trying to reach an optimal mean pairwise identity around 80%.

The default *behaviour* can be customized in various ways (use —help for details). The following command, for example, samples three different alignments with four sequences each.

```
# rnazSelectSegs.pl--num-segs=4 \
--num-samples=3 miRNA.maf|RNAz
```

By default, the first sequence in the alignment is always in the set of selected sequences. This is the desired behaviour for genomic screens, where one usually likes to retain a reference sequence.

Developing Long Alignments

RNAz cannot score alignments longer than 400 columns. In practice, it is generally advisable for the user to score long alignments, say >200 columns, in shorter, *overlapping windows*. For general *purpose screens,* we recommend a window size of 120. This window size appears large enough to detect local *secondary* structures within long ncRNAs and, on the *other hand,* small enough to find short secondary structures without *loosing* the signal in a much too long window.

The file unknown.aln contains a *noncoding* region conserved in vertebrates. The file can be scanned for RNA secondary structures by typing:

```
# rnazWindow.pl--window=120--slide=40 unknown.aln\
| RNAz--both
```

The results show that RNAz does not predict an RNA in this region. On UNIX like system, "*/ grep Prediction*" can be added to get a quick overview on the results. The *rpazwindow. p1* program has numerous additional functions.

Genomic Screens of Commercial Level

In the following, we demonstrate the usage of this pipeline on the example of a genomic

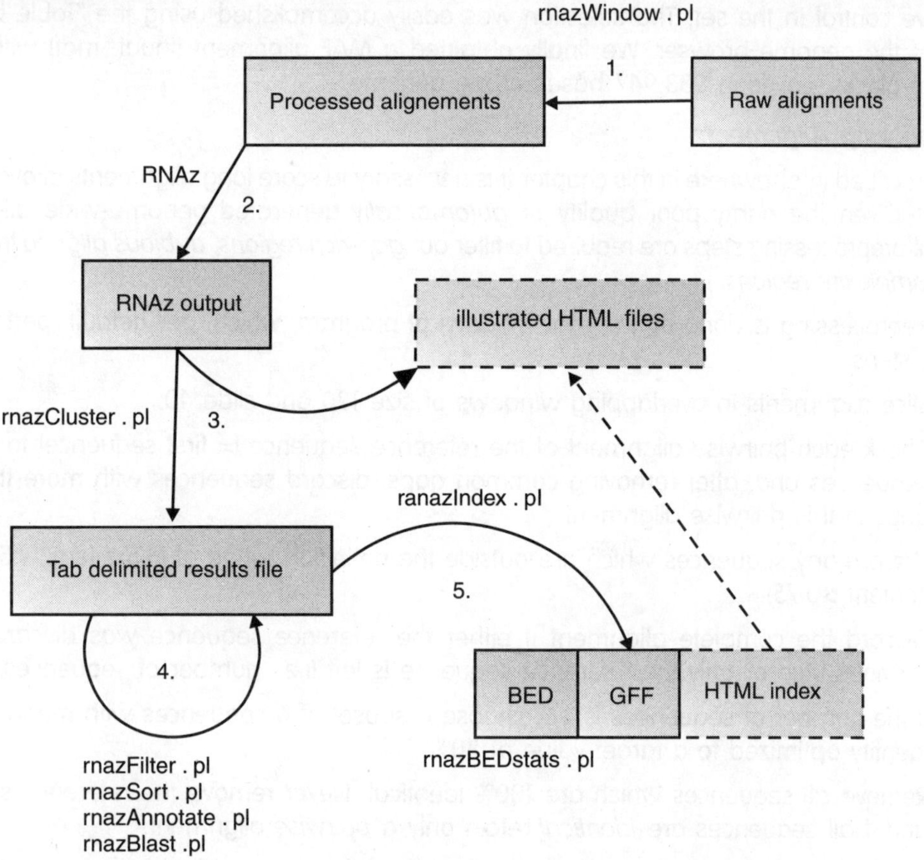

Figure 13.2: Showing the RNAz and the helper programs.

screen of *S. cerevisiae*. We want to describe the method as general as *possible* and we will focus here mainly on technical details. A paper describing the results of a *comprehensive* RNAz screen in yeast is in *preparation*.

Reasonable Choosing

Choosing a reasonable set of input alignments is one of the most important steps during the analysis. There are a variety of different programs available to generate genome-wide alignments. Here, we use Multiz alignments of up to seven *Saccharomyces* species, which can be downloaded from the UCSC genome browser (genome.ucsc.edu). In principle, we could use all *alignments* covering the complete genome.

The biggest problem in large genomic screens is probably *specificity*. We have a *relatively* constant background signal of *false-positives*. The more sequences we put into the screen, the more false-positives we get out. It is, therefore, a good idea to choose the input set as small as possible (trying not to discard any *interesting* regions of course).

In our case, we only analyze the *intergenic* regions, i.e., we discard any *coding regions* and all other annotated features (*pseudogenes*, repeats, ARS elements,...). We retain known ncRNAs as positive control in the set. The selection was easily accomplished using the "Table browser" feature of the genome browser. We finally obtained a MAF alignment (input. maf) with 10,822 alignment blocks, covering 983,947 bases of the genome.

Observing long Alignments

As described in elsewhere in this chapter it is necessary to score long alignments in overlapping windows. Given the partly poor quality of *automatically* generated genome-wide alignments additional preprocessing steps are required to filter out *gap-rich regions, dubious aligned* fragments or low *complexity regions.*

All preprocessing is done by the rnazWindowl pl program, which, per default, performs the following steps:

1. Slice alignments in overlapping windows of size 120 and slide 40.
2. Check each pairwise alignment of the reference sequence (= first sequence) to all other sequences and, after removing common gaps, discard sequences with more than 25% gaps in this pairwise alignment.
3. *Discard* any sequences which are outside the definition range of RNAz (e.g., <50 nt, GC content >0.75).
4. Discard the complete alignment if either the reference sequence was discarded in a previous step or only the reference sequence is left (i.e., number of sequences <2).
5. If the number of sequences is >6, choose a subset of 6 sequences with mean pairwise identity optimized to a target value of 80%.
6. Remove all sequences which are 100% identical. Never remove the reference sequence and if all sequences are *identical* retain only a *pairwise* alignment.

All these steps can be *customized* with the *appropriate* command-line parameters. Here, we use the default settings. We define, however, a *minimum number* of four sequences in the *alignment retaining* only regions that are well conserved across several species:

```
# rnazWindow.pl  —min-seqs=4  input.maf  >windows.maf
```

This command will take a few minutes.

Running RNAz

The file windows. maf is now ready for being scored with RNAz. We use the `--both-strands` parameter to score both the forward and the reverse *complement* strand. We also `set--show-gaps`, which means that the output is shown *including* the gaps. With this option it is possible to recover the complete alignment from the RNAz output file, which is useful in later steps of the pipeline. Finally, we set a p-value cutoff of 0.5, meaning that only *positive predictions* are stored resulting in a much smaller output file.

```
# RNAz--both-strands--show-gaps--cutoff =0.5\
windows.maf > rnaz.out
```

This will take approx 1 h on a modern desktop computer but may vary depending „on the system.

Placing the Results

The file rnaz. out now holds all windows that have a positive RNAz signal with $p > 0.5$. It is possible that several windows cover the same genomic region. Overlapping windows are therefore clustered in *loci:*

```
# rnazCluster.plrnaz.out>results.dat
```

This command assigns each window a consecutively numbered "window ID" and each group of overlapping windows a "locus ID." For each window and each locus all relevant data (use `--help` for details) is stored in a tabulator separated text file.

Inspecting the file results. dat, we see that we have 1104 windows which can be grouped in 454 loci.

It is important to note that the term "*locus*" must not be understood in the sense of a genetic unit. It is, of course, possible that several loci of our procedure cover one long ncRNA gene.

At this point we also want to add that we are painfully aware of the fact that the process of first slicing the *alignments* and the *reclustering* of them is not optimal. Ideally, one would like to *predict* conserved RNA structures locally without sliding windows. Although this should be possible and we are working on a local version of RNAz, the sliding window approach is currently the only *reasonable protocol.*

Sorting the Results

The data file now contains the raw data of all hits. In the following analysis steps, one usually wants to filter and sort candidates by various criteria. For this purpose, the programs rnazFilter. pl and rnazSort. pl can be used. For example,

```
# rnazFilter.pl  "P>0.9"  results.dat
```

lists all windows that have a p-value more than 0.9. For hints on how to formulate more complex filtering expressions see Note elsewhere in this chapter. The `--count` option will count the hits. We have 670 Windows in 303 loci on the $p > 0.9$ significance level. In addition, we can sort the hits:

```
# rnazFilter.pl"P>0.9" results.dat\
| rnazSort.pl combPerPair
```

This sorts the output by the "Combinations/Pair" value, i.e., by compensatory mutations supporting the structure.

Sub-Selection

Using different combinations of `rnazFilter.pl` and `rnazSort.pl`, various subselections of the complete data from results. dat can be created. There will always be a *tabulator* delimited data file. The program `rnazIndex.pl` helps to convert these kind of data files into the standard *annotation* formats GFF (`--gff`) or BED (`--bed`).

GFF (http://www.sanger.ac.uk/Software/formats/GFF/) is a widely used format supported by many programs. BED (http://genome.ucsc.edu/FAQ/FAQformat) is the native annotation format for the UCSC genome browser but is *generally* useful because of its simplicity (in its simplest form it is a list of genomic *locations*: sequenceID start stop).

The following command creates a GFF file from all results:

```
# rnazIndex.pl--gff results.dat>results. gff
```

Observing the Results on a Website

It is often insightful to manually check individual predictions, for example by analyzing different illustrations of consensus structures. The *creation* of the necessary files is a tedious task, which, however, can easily be *automatized*. If the cluster command from elsewhere in this chapter is run with the option `--html`,

```
# rnazCluster.pl--html  rnaz.out>results.dat
```

the program generates image files for all hits. For the `--html` option to work, the Vienna RNA package must be installed (including the Perl programs of the Utils directory) and the program *Ghostscript*, rnazCluster. pl creates a subdirectory called results, which, in turn, has a subdirectory locusN for each locus. In the locusN directories, the images files can be found together with an index. html, which arranges the images for each locus on a webpage. The index files can be opened using almost any web browser.

To get an HTML-formatted table of all hits linking to the subpages for each locus, use rnazIndex.pl with the —html option:

```
# rnazIndex.pl--html results.dat>results/results.html
```

Additional Annotation

Once there is a list of predicted RNAs, the user may want to add additional *annotations* to the predictions. Additional fields can be added to the *tabulator* separated data file. Here, we demonstrate this by *comparing* our prediction with the known ncRNA *annotation* from the *Saccharomyces* genome database. The program `rnazAnnotate.pl` checks each predicted locus for overlap with an annotation file in BED format:

```
# rnazAnnotate.pl --bed../sgdRNA. bed results.dat\
> annotated.dat
```

We find that out of 454 predicted loci, 280 overlap with known ncRNAs (of the 303 loci with $p > 0.9$, 215 are known ncRNAs). We detect all sorts of different ncRNA classes (tRNAs, rRNAs, snRNAs, snoRNAs, RUFs, and other ncRNAs like telomerase RNA or RNAseP,...). Most of the known 373 ncRNAs in yeast are tRNAs (275), which are partly difficult to detect in this screen because most of them are ti 100% conserved (i.e., no covariance information).

Without providing a detailed sensitivity analysis for this specific yeast screen, we want to add that sensitivity highly depends on the ncRNA class. MicroRNAs, for example are easy to detect because of the high thermodynamic stability of the hairpin precursor.

On the other hand, C/D type snoRNAs for example are generally difficult to detect because they lack a pronounced secondary structure. We completely miss ncRNAs, which do not depend on a secondary structure for their function, as for example the yeast *SER3* regulating RNA, which, as expected, does not show up in this screen.

Databases to Know ncRNAs

Another possibility to annotate predicted ncRNAs is to compare the sequences to databases of known ncRNAs. In the following, we match the predicted loci against the Rfam database using a simple Blast sequence search. Alternatively, one could use more sensitive methods, which also incorporate secondary structural information (e.g., Infernal $_k$To run this example, the *S. cerevisiae* sequence files will be needed, the Rfam database file and a working NCBI Blast installation.

First, change into the directory rf am and run:

```
# formatdb -t rf amT-i rfam -p F
```

This command creates the index files for the file rf am, which is a Fasta formatted file with all entries of the database. Now run:

```
# rnazBlast.pl--database rfam --seq-dir=seq\
--blast-dir=rfam results.dat>annotated.dat
```

This program takes the *S. cerevisiae* reference sequence for each locus and runs a Blast search against the Rfam database. If there is a hit with an expectation value below some cutoff (default: $E < 10^{-6}$), the name of the matching database query is added as a new field to the data file. Please note that the locations of the sequence data files and the blast index files on the command-line must be specified.

Use of Statistics

To get an impression of the false-positive rate of a specific screen it is useful to do a control screen on randomized alignments. The command

```
# rnazRandomizeAln.pl input.maf>random-input.maf
```

will produce a randomized version of the input alignments by shuffling the positions in the alignments. The program aims to remove any correlations arising from a natural *secondary structure* while preserving important *alignment* and sequence characteristics as for example mean pairwise identity or base *composition.*

We repeated the complete analysis with the randomized alignments and we get 102 and 39 loci, on the $p > 0.5$ and $p > 0.9$ level, respectively.

Table elsewhere in this chapter summarizes all results of this example screen. There are a few programs that will help with the gathering statistics on the data. For example,

```
# rnazIndex.pl --bed results.dat \
|rnazBEDsort.pl|irnazBEDstats.pl
```

gives detailed information on the *predicted loci,* including the covered genomic region in nucleotides. This command first exports the results as BED file, *sorts* the results by the genomic *location* and, finally, evaluates the coordi-nates in the BED file. To get statistics on the input alignments, use a command like this:

```
# rnazMAF2BED.pl --seq-id=sacCer windows.maf\
| rnazBEDsort.plirnazBEDstats.pl
```

Table 13.1: Statistics of the Yeast Example Screen.

	p > 0.5	p > 0.9
Predicted loci	454	303
Known ncRNAs	280	215
Loci without annotation	174	88
Predicted bases	60,834	44,082
Fraction of input alignments (%)	10.6	7.7
Predicted loci random	102	39
Predicted bases random	12,823	6017
Fraction of input alignments random (%)	2.2	1.0

`rrnazMAF2BED.pl` converts a MAF formatted alignment file to coordinates in BED format. With `--seq-id` the user specifies which sequence is used as reference.

Using these tools, the user will find, for example, that in the random control 1.0 % of the input sequences are predicted as R)TA on the $p > 0.9$ level. This is exactly the false-positive rate as expected. The absolute number of *false-positives*, however, strongly depends on the specific screen. In this example we have 88 hits p > 0.9 without RNA *annotation* and find that 39 hits should be expected by chance.

So we must expect that roughly half of our predictions are false-positives. On the other hand, this implies that the other half of the predicted loci should be real functional RNA structures, either as part of a ncRNA or as regulatory element of a mRNA. However, one always have to bear in mind *possible shortcomings* of this kind of *random* control.

PONTS TO REMEMBER

1. *Custom installation of RNAz*

The installation process using ./configure and make should work on all UNIX-like systems. If there is an error messages, it may be necessary to install additional *"developer packages."* On some Linux distributions, for example, there is no C-compiler installed by default. Also, on OS X it is necessary to install the "XCode" tools.

If the user does not have root privileges or wants to install RNAz into a different location than /usr/local/ (e.g., the home directory) the following command can be used:

```
#./configure--prefix=/home/stefan\
—datadir=/home/stefan/share
```

This installs the executable to /home/stefan/bin and the example files, Perl programs, and other data to /home/stefan/share/RNAz. Please note that the bin directory must be in the PATH of executables if they are to be called the RNAz executable without specifying the complete path.

2. *Running the Perl programs*

Because dt'lferent people usually like to have their scripts in different locations, the Perl programs are not installed to `/usr/local/bin` by default. They are installed to `/usrlocal/share/RNAz/perl`. To make them available from other locations, copy all files from this directory to a directory, which is included in the PATH of executables, e.g.:

```
#cp/usr/local/share/RNAz/perl/*/usr/local/bin
```

Alternatively, the directory with the Perl programs can be added to the PATH variable by editing the . bashrc or . cshrc file in the home directory. In any case, it is important that the Perl *module* file RNAz. pm resides in the same directory as the Perl programs (`*.pl`). All the Perl programs depend on this module file.

Another important point is that the Perl programs expect that the path of the Perl executable is `/usr/bin/perl`. This is the standard location on almost all Linux/UNIX systems and OS X. If the Perl installation is different, the user has to customize the first line of all the Perl programs according to the location of the perl executable.

On a Windows system the Perl programs should work if Perl has been installed. and the Path variable has been set.

3. Optional software on Windows on OS X. It might be a bit tricky to install the necessary software on Windows and OS X for the programs.

The Vienna RNA package and NCBI Blast can be installed on OS X without problems. However, unlike on a Linux system, Ghostscript is not installed per default. A *precompiled* package can be obtained from fink. sourceforge. net or *darwinports*.opendarwin.org.

Alternatively, the source can downloaded from http://www.ghostscript.coml and the package can be built with . /configure and make.

Ghostscript can be installed on Windows through a simple installer file that can be downloaded from http://www.ghostscript.com/. Follow the installation instruc-tions. Locate the newly installed file gswin32c.exe and copy it to a folder, which is in the Path (e.g., the folder, where the RNAz . exe executable resides). Rename the file to gs . exe.

Windows users do not have to install the Vienna RNA package. The relevant programs are part of the RNAz Windows installer.

To install NCBI blast on windows, create a new folder (e.g., c : \Program Files\ blast) and download the blast-2. *-win32 . exe file from ftp://ftp.ncbi.nih.gov/ blast. Within the new folder, double click on the blast-2 . *- win32 .exe file, which extracts the programs and data. Add the bin subdirectory to the Path: right-click "My computer", then click "Properties". Select "Advanced/ Environment variables/New". Add the complete path of the blast bin directory to the variable Path, use "•" s separator.

4. *Creating the input alignments*

RNAz can only detect a conserved structure if This structure is accurately reflected in the alignment. Therefore, the quality of the alignment is crucial for the success of the analysis. In practice, we found that if the alignment has a mean pairwise identity more than approx. 60% simple sequence-based progressive, *global alignment* methods yield *reasonable* results and there is not much difference between methods. One of the best programs for aligning RNAs is Clustal W.

For genome-wide alignments we have only experience with Multiz alignments. Also these alignments are of reasonable quality and there is generally no need for realignment. We suppose that also other genome-wide alignment methods produce suitable alignments as long the aligned regions are of *sufficient similarity* (mean pairwise identity somewhere around 60% or higher). In cases with sequences less than 60% identity, simple sequence-based methods usually do not find an optimal structural alignment.

Although, in principle, structural enhanced alignments could help here, this alternative is not relevant in practice. First, there are hardly any structural multiple sequence alignment programs available. Second, current approaches are much too slow to use them for every day analysis. Third, RNAz is not trained on structural alignments. In contrast to pure sequence based alignment, the user would get unusual high SCIs. This could confuse the decision model and the user would get unpredictable results.

5. *Additional values*

The consensus MFE, which is calculated by the RNAal-ifold algorithm. can be split in two terms. One is the "energy contribution," which is the folding energy from the standard energy model. The "*covariance contribution*" is the part that comes from the additional "*bonus*" or "*penalty*" energies for *compensatory/consistent* and *inconsistent* mutations, respec-tively. If the covariance term is negative, there are more compensatory mutations than inconsistent mutations.

RNAz also calculates another value quantifying compensatory/consistent mutations: "Combinations/Pair." This is the number of different basepair combinations in the consensus structure divided by the number of pairs in the consensus structure. Both the covariance contribution of the consensus MFE and the "Combi-nations/Pair" are mainly useful for final sorting a set of equally good predictions that have been filtered using other criteria (e.g., *p*- or *z*-scores).

RNAz uses a SVM algorithm for classification. The raw output of the SVM is the so-called "decision-value." This real-valued number is positive if the prediction is "RNA" and negative otherwise. From this value we calculate the more intuitive "RNA class probability" or "*p*-value," which is 0.5 for a decision value of 0. In some cases, the raw decision value can be more convenient than the *p*-value (e.g., to plot the distribution of RNAz results).

6. *Estimating flase-positives*

The RNAz classification model is trained on a test set consisting of natural RNAs as positive examples and randomly shuffled alignments as negative examples. Thus, any signal reported by RNAz is relative to an artificial background. Although this null model of shuffled sequences is probably the most sensible choice possible, one cannot assume that it behaves exactly like the natural background of real sequence data.

Also the estimation of false-positive rates is based on shuffled sequences. We want to stress that, therefore, such an estimation of false-positives must be regarded as a lower bond because one cannot rule out the possibility that nonrandom patterns in natural sequences cause a higher rate of false-positives than one observes in synthetic random sequences. In particular, the z-score calculation might be affected by such effects. For example di-nucleotide content could bias the MFE structure prediction.

As an opposite effect one must consider the possibility that the shuffling procedure cannot remove all secondary structure signals and, thus, overestimates the real false-positive rate. If an alignment is shuffled with many compensatory mutations, the number of "compatible columns" remains the same, allowing for compensatory mutations also in the shuffled alignment.

7. *Manual inspection of candidates*

If the user has a hit with p > 0.9, there is aproximately a chance of 1 in 100, that this arises through pure chance. It makes sense to critically look at a hit. Sometimes the signal only comes from a low *z*-score of borderline significance and there is no evidence for structural conservation. Sometimes the complete alignment looks pathological (weird gap patterns, low complexity regions, and so on), which suggests that this is not a relevant structure. It is useful to analyze a predicted structure with RNAalifold and its visualization methods .

Visual inspection of a color-coded alignment and the consensus structure gives an idea about compensatory mutations supporting the structure and inconsistent mutations, which do not support the structure. It must be noted that many ncRNAs in real life-data are not supported by compensatory mutations, still they can be detected based on the stability or the SCI. The SCI implicitly also considers the mutational pattern outside of stems. To conclude, the p-value efficiently filters the data for candidates, but only the complete picture can help in the decision on the relevance of a hit.

8. *Advanced filtering*

Filtering the tab-delimited data files using standard UNIX tools like grep or awk is difficult because of the special window/locus grouping of the data. The rnazFilter. pl program can be used. The filter statement uses the field names (e.g., z, SCI, combPerPair, *see*—help for a complete list) and standard logical operators as used in the Perl language: > (greater than), < (smaller than), = = (equals numerically), eq (equals string), not, and, or, = ~/regex/ (pattern match). In addition, brackets can be used to group and combine statements. For example the following statement gives all windows with p > 0.9 and *z* < –3 on chromosome 13:

```
# rnazFilter.pl "P>0.9 and z<-3 and segID=~/chr13/"\results.dat
```

It is important that everything put in the filter statement is evaluated by the Perl interpreter. This can be potentially harmful, so take care.

Chapter 14

cDNA Macroarray from *Arabidopsis thalinana*

Recently, measurements of gene *expression* levels have been carried out on a large scale by using both cDNA microarray and cDNA *macroarray technology*. cDNA expression arrays generally consist of a nylon membrane (cDNA macroarray) or glass slides (cDNA microarray) spotted with polymerase chain reaction (PCR) products produced from cDNA clones. The arrays are *hybridized* with *radioactively* or *fluorescently* labeled probes *transcribed* from rRNA.

Both types of cDNA array have become very useful in the elucidation of mRNA expression *patterns* among different tissues, the determination of *differential* gene expression during growth or as the result of *environmental* conditions, and the identification of novel genes. Apart from the high number of genes screened, an advantage of cDNA *microarrays* is the sensitivity of detection because of the high probe *concentration*, which is a result of the small *incubation* volumes used.

However, in a comparative study, cDNA macroarrays used in conjunction with 33p labeled probes showed higher *sensitivity* than cDNA microarrays because of the larger amount of DNA spotted on the membrane. Given the generally held opinion that 300,000 mRNAs are expressed per cell, an mRNA whose expression level is only 0.005% (i.e., 15 molecules per cell) *theoretically* can be detected using cDNA *macroarrays*. Currently, the cDNA macroarray method is more accessible to *researchers* because, unlike *microarray technology*, it can be carried out using standard laboratory equipment.

However, a disadvantage of cDNA macroarrays is that only one sample can be analyzed per membrane, resulting in variations in signal intensity because of the *potential* presence of different amounts of eDNA probe when comparing two or more samples. In contrast, cDNA *microarrays* on *glass slides* allow for direct *comparison* of two samples in the same array because each sample is labeled with a different *fluorescent* probe, such as Cy3 or Cy5.

This practice allows accurate measurement of each signal. Therefore, a 2-fold up- or down-regulated gene can be identified using a single cDNA *microarray*. Contrastingly, *comparative* experiments using cDNA macroarray are easily affected by experimental *variations*. Consequently, statistical analysis of data from replicate *experiments* is needed to reduce the effects of such *variations*.

Table 14.1: Comparison of cDNA macroarray and cDNA microarray.

	cDNA microarray	cDNA macroarray
Support	Slide glass	Nylon membrane
Degree of location	< 500 genes/cm^2	500-10,000 genes/cm^2
Detection system	Radioisotope (^{33}P)	Fluorescent dye (Cy3, Cy5)
Possibility of reuse	Yes	no
Sensitivity	0.005% mRNA	0.05% mRNA
Dynamic range	10^4-fold	10^2-fold
Deviation	20%	100%

REQUIREMENTS

Sampling

1. Culture medium: 1:2 (v/v) Gamborg B5 Complete medium (GibcoBRL, Rockville, MD), 1:1000 (v/v) Hyponex (Hyponex 10:5:10; Hyponex Japan, Osaka, Japan), 1 % sucrose; pH adjusted to 5.7.

Laboratory Preparation

1. 4 *M guanidium thiocyanate* buffer: 4 *M guanidium thiocyanate*, 0.1 *M* Tris-HCl, pH 7.5, 10 mM EDTA, 0.5% Sarkosyl, 0.1% (3-mercaptoethanol.
2. Phenol-chloroform: 50% (v/v) liquid phenol, pH 8.0, 48% (v/v) chloroform, 2% (v/v) isoamyl alcohol.
3. 5.7 *M* CsCl cushion: 5.7 *M* CsCl, 0.1 *M* EDTA; pH adjusted to 7.5.
4. DEPC-treated water: 0.1 % diethyl pyrocarbonate.
5. 10× Hybridization buffer: 1.2 *M* NaC1, 0.1 *M* Tris-HCl, pH 8.0, 50 mM EDTA, 10% sodium dodecyl sulfate (SDS).
6. Array BPB solution: 10 mM Tris-HCl, (pH 7.5, 0.25% bromophenol blue, I *mM* EDTA, pH 8.0, 60% (v/v) glycerol.
7. Alkaline solution: 0.5 *N* NaOH, 1.5 *M* NaCl.
8. Neutralizing solution: 1.5 *M NaCl,* 0.5 *M* Tris-HCl; pH adjusted to 7.4.
9. 2×SSC: 300 *mM* NaCl, 30 mM sodium citrate.

Method of Hybridization

1. 5× M-MLV RT buffer: 250 mM Tris-HCl, pH 8.3, 375 mM KCl, 15 mMMgCl$_2$.
2. dNTP mixture without dCTP: 20 mM dATP, 20 mM dTTP, 20 mM dGTP.
3. Hybridization solution (Church phosphate buffer): 0.5 MNa$_2$HPO$_4$, 1 MM EDTA, 1 μg/mL.

poly dA [Roche Diagnostics, Indianapolis, IN], 7% SDS. Adjust pH to 7.2.

4. 0.2× SSC: 30 mM NaCl, 3 mM sodium citrate.

PROCEDURES

Sampling

1. *Arabidopsis thaliana* (*L.*) Heyhn accession Columbia (Col-0; ABRC, Columbus, OH) is used for analysis.
2. Plants for constructing cDNA library are grown in soil under a 16:8-h light:dark cycle at 22°C.
3. Above ground organs, flower buds, and green siliques are harvested from 2- to 6-week-old plants. For liquid culture, sterile seeds are sown in culture *medium* and *grown* for 2 weeks under *continuous* light at 22°C with rotation. *Seedlings* and roots are collected from *liquid-cultured* plants.
4. For isolation of mRNA, plants are *germinated* on blocks of rock wool and grown in a growth chamber at 25°C at a relative *humidity* of 50-60% under a photosynthetic *photon* flux density (PPFD) of 100 μmol m^{-2} s^{-1} in 14:10-h light:dark cycles.
5. Plants are watered with a 5:10:5 liquid fertilizer (Hyponex, Hyponex Japan) diluted 2000-fold. Two-week-old plants are exposed to a single dose of 200 nL/L O_3 for 12 h in an O_3 chamber, as described previously. The O_3 chamber is maintained at 25°C and a relative humidity of 70% under a PPFD of 100 μcool m^{-2} s^{-1} in *continuous* light. Plants *remaining* in *ambient air* serve as the controls. O_3 is generated with an O_3 generator (Sumitomo Seika Chemicals, Tokyo, Japan).

Laboratory Preparation

Method of Isolation of Poly(A)+ RNA

1. Total RNA is extracted from above ground organs, flower buds, roots, 2- to 6-wk-old plants, and liquid-cultured seedlings by the guanidium thiocyanate-CsCI ultracentrifugation method.
2. Four grams of frozen tissue are ground to powder in liquid nitrogen and treated with 25 mL of 4 *M* guanidium thiocyanate buffer.
3. Centrifuge the homogenate at 5,000g for 10 min at room temperature.
4. Transfer the supernatant to a fresh tube and add an equal volume of phenol-chloroform. Mix well.
5. Centrifuge the mixture at 5,000g for 5 min at room temperature.
6. Transfer the supernatant to a fresh tube, and repeat steps 4 and 5 four times.
7. Layer the supernatant onto a 15-mL 5.7 *M* CsCI cushion in a clear ultracentrifuge tube and centrifuge for 22 h at 20°C and 271,900g using a RP50VF rotor (Hitachi Koki, Tokyo, Japan).
8. Invert the tube to discard the supernatant, and allow any remaining fluid to drain into the

paper towel.

9. Wash the pellet of RNA with 70% ethanol at room temperature. Invert the tube and drain off the ethanol.
10. Allow the pellet of RNA to dry, and dissolve it in DEPC-treated water.
11. Poly(A)$^+$ RNA is purified from total RNA by using an mRNA purification kit according to the manufacturer's instructions (Amersham Biosciences, Piscataway, NJ).

cDNA Library

1. 1.5 to 7.5 μg poly(A)$^+$ RNA is primed with oligo(dT)^{18} primer carrying an *XhoI* site, and the first-strand synthesis reaction is performed for 30 min at 52°C using SuperScript II (Invitrogen, Groningen, Netherlands) reverse transcriptase according to the manufacturer's instructions.
2. After second-strand synthesis, cDNA is blunt-ended with using DNA Blunting Kit (TaKaRa Bio, Ohtsu, Japan), and *EcoRI* adaptors are ligated to both ends.
3. The *XhoI* sites at the 3' ends of the cDNA are generated by digestion with *XhoI.*
4. Size selection of cDNAs before vector ligation is performed as follows. Synthesized cDNAs are separated by 1 % agarose gel electrophoresis, and fractions ranging from 1 to 3 kb are recovered from the gel using a QIAquick Gel Extraction Kit (QIAGEN, Dusseldorf, Germany) as described by the manufacturer's instructions.
5. The recovered fragments are cloned into the *EcoRI-XhoI* sites of pBluescript II SK$^-$ plasmid vector (Stratagene, La Jolla, CA) and transformed into *Escherichia coli (E. coli)* XLI-Blue MRF' strain (Stratagene) by electroporation.

Normalization of cDNA Library

1. For normalization of the library, a single-stranded library is prepared from 5 μg of plasmid DNA. First, a single-stranded nick is introduced into the plasmid DNA using 20 U of gene II endonuclease of phage F 1 (M13 gpII protein; Life Technologies, Gaithersburg, MD) for 30 min at 37°C. This enzyme makes a nick at the fl replication origin in the plasmid DNA. Then, a single-stranded plasmid is generated with digestion of 250 U of exonuclease III (New England Biolabs) that has 3'to 5' exonuclease activity. The reaction is performed using attached buffer (TaKaRa Bio, 50 mM Tris-HCl, pH 8.0, 5 mM MgCl, 10 mM 2-mercaptoethanol), and the volume adjusted to 50 μL with water. Then incubate the reaction mixture for 1 h at 37°C.
2. The cDNA inserts are allowed to self-hybridize by using an excess (about 1 μg) of cDNA inserts generated by PCR amplification. For amplification of the inserts, about 5 ng of DNA template (the single-stranded library plasmid produced described elsewhere in this chapter) was mixed with 1 μL of 100 μ*M* T7 primer. 5'-TAATACGACTCACTATAGGG-3' and 1 μL of 100 μ*M* SK primer 5'-GCTCTAGAACTAGTGGATC-3', and then PCR was performed using *Ex-Taq* DNA polymerase (TaKaRa Bio) in a Perkin-Elmer 9600 Thermal Cycler (Applied Biosystems, Foster City, CA) according to the following program: 7 min ramp from room temperature to 94°C; 20 cycles of 1 min at 94°C, 2 min at 55°C, and 3 min at 72°C. A final extension of 7 min at 72°C follows.

3. The PCR product is ethanol-precipitated, dissolved in 1.5 µL of water, and then mixed with 5 µL of the single-stranded library cDNA (50 ng) dissolved with formamide, 0.5 µL(10 µg) of the 5'-blocking oligo nucleotide mixture. (5'-GCTCTAGAACTAGTGGATCCCCC GGGCTGCAGGAATTCG-3' and 5'-AATTCGGCACGAG-3'), and 0.5 µL (10 µg) of the 3'-blocking oligo nucleotide mixture (5'-CTCGAGGGGGGGCCCGGTA-3' and 5'-GTA CCCAATTCGCCCT ATAGTGAGTCGTATTA-3').
4. The mixture is kept at 80°C for 3 min, 1 µL of 10× hybridization buffer and 1.5 µL of water are added, and hybridization is performed at 30°C for 24 h.
5. The remaining single-stranded circles are purified using hydroxyapatite chromatog-raphy and converted to double strands using the Klenow fragment (TaKaRa Bio). The double-stranded DNA obtained is transformed into host E. *coli* by electroporation. The normalized library typically contains 1×10^6 independent clones.

Sequencing

1. Plasmid DNA is prepared by the alkaline lysis method in a 96-well plate.
2. Clones are digested with *ApaI* and *SmaI* to release the inserts, whose sizes are measured by agarose gel electrophoresis of the digestion reactions.
3. Clones undergo nucleic acid sequencing with a BigDye Terminator Cycle Sequencing Ready Reaction Kit (Applied Biosystems) according to the manufac-turer's instructions and are electrophoresed using automated DNA sequencers (ABI PRISM 373 and 377XL; Applied Biosystems).
4. Random cDNA clones are chosen from the normalized cDNA libraries constructed. In the example experiment, 14,026 clones in total were sequenced from the 5'-end, and 39,207 clones were sequenced from the 3'-end. As a result, 39,207 3'-expressed sequence tag (EST) sequences were clustered into 12,028 independent groups.
5. 12,028 independent EST clones were used for construction of a massive cDNA macroarray filter. Inserts of EST clones were amplified using 5'-GTAATACGACTCACTATAGGGC-3' and 5'-TCATTAGGCACCCCAGGC TTTACAC-3' as the primers. PCR was performed using a Perkin-Elmer 9600 Thermal Cycler (Applied Biosystems) in 80-µL reaction mixtures, each containing 1 µL of cultured E. *coli* cells as the PCR template, lx *Ex-Tay* Buffer (TaKaRa Bio), 2.5 m.*M*$MgC1_2$, 0.2 mM dNTPs, 0.2 pM each primer, and 0.5 U of *Ex-Tay* DNA polymerase (TaKaRa Bio). Amplification conditions are 35 cycles of denat-uration at 94°C for 45 s, annealing for 45 s at 55°C, and elongation at 72°C for 2 min; the final extension is for 4 min at 72°C.
6. Five microliters of each PCR-amplified product is electrophoresed through a 1.5% agarose gel to check amplification of the EST fragment.

cDNA Macroarray Filters

1. The remaining 75 µL of PCR product is mixed with 15 µL of array BPB solution and spotted onto an 8 × 12 cm nylon filter (Biodyne A, Pall, East Hills, NY) in duplicate by using a Biomek 2000 Laboratory Automation Workstation (Beckman Instruments Inc., Fullerton, CA). This pin blotter spots 2,880 EST clones on a filter. Therefore, for the example experiment, five independent filters were made to accommodate the 12,028 independent EST clones.

An example cDNA macroarray filter is shown in Figure elsewhere in this chapter.

2. λ DNA (10 mg/µL) is spotted in duplicate as negative controls.
3. Spotted filters are denatured in a shallow tray of alkaline solution for 2 min, after which the filters are transferred into neutralizing solution for 2 min.
4. Rinse the filter in 2× SSC and place it, DNA side up, on paper towels to dry.
5. After the filters are dry, the spotted ESTs are fixed on the filters by UV cross linking (1.2 J/cm^2).

Method of Hybridization

RNA Labeling

1. Total RNA or mRNA that is isolated from sample of interest is prepared as described earlier. For the example experiment, total RNA was isolated from O_3-exposed and -unexposed *Arabidopsis* plants.
2. Ten micrograms of total RNA (or 0.2 µg of poly(A)$^+$ RNA, if available) is mixed with 2 pg of oligo-dT$_{(12-18)}$ (Invitrogen) and the volume adjusted to 14 µL with DEPC-treated water.
3. Incubate the RNA solution for 10 min at 70°C. Chill the sample on ice for 2 min.
4. Add 6 µL of 5x M-MLV RT buffer, 1.5 µL of dNTP mixture without dCTP, 1 µL of 0.1 M dithiothreitol, 6 pL of [a*-^{33}P] dCTP (spec act > 2,500 Ci/mmol; Amersham Biosciences), and 7.5 U SuperScript II (Invitrogen).
5. Mix well and incubate the reaction mixture at 37°C for 90 min. Chill the sample on ice for 5 min.
6. Apply the sample to a G-50 spin column (Probe Quant G-50 Micro Column; Amersham Bioscience). Centrifuge at 820g for 2 mina
7. Collect the effluent (about 30 pL) and denature it at 95°C for 5 min.
8. Keep the denatured probe on ice until it is used.

Laboratory Preparation

1. Place the cDNA macroarray filters into a hybridization bag (Atto, Tokyo, Japan).
2. Pour in 10 mL of hybridization solution into the hybridization bag and then seal it with a heat-sealer.
3. Submerge the filters in a water bath at 65°C and incubate for 1-2 h.
4. Remove the bag containing the filters from the water bath. Open the bag by cutting off one corner with scissors.
5. Add the denatured probe to the hybridization solution and then squeeze as much air as possible from the bag.
6. Reseal the bag with the heat-sealer.
7. Submerge the bag in a 65°C water bath for more than 16 h.
8. After hybridization, open the bag by cutting off a corner with scissors.

9. Transfer the filters to a flat-bottom plastic box containing 250 mL of 0.2× SSC, 0. 1% SDS.
10. Wash the filters for 15 min at 65°C with gentle agitation.
11. Replace the solution with fresh 250 mL of 0.2× SSC, 0.1% SDS, and transfer the box to a water bath set at 65°C with gentle agitation for 15 min.
12. Remove most of the liquid from the filters by placing them on paper towels.
13. Cover the filters with plastic wrap and expose them to a bioimaging plate (Fuji Film, Tokyo, Japan) for more than 12 h.

Mathematical Calculations

1. Scan autoradiographs on a high-resolution scanner (Storm; Amersham Biosciences) with 50-pm resolution, and quantify the signal intensity with Array Vision software (Amersham Biosciences) as described in the manufacturer's instructions. An example scanned image is shown in Figure elsewhere in this chapter.
2. Export the quantified signal intensity as a '.csv' format file and import the file into a data analysis program (e.g., Microsoft Excel Version X, Microsoft, Redmond, WA) for analysis.
3. Normalize interfilter differences in signal intensity by adopting global normal-ization in the following way. Average the intensities of all signals in the filter, and then calculate the relative signal intensity as the ratio of the signal of interest to the average intensity of each filter. Thus the estimated values are designated as expression ratios.
4. Before further analysis, average the expression ratios of the duplicate spots. To select

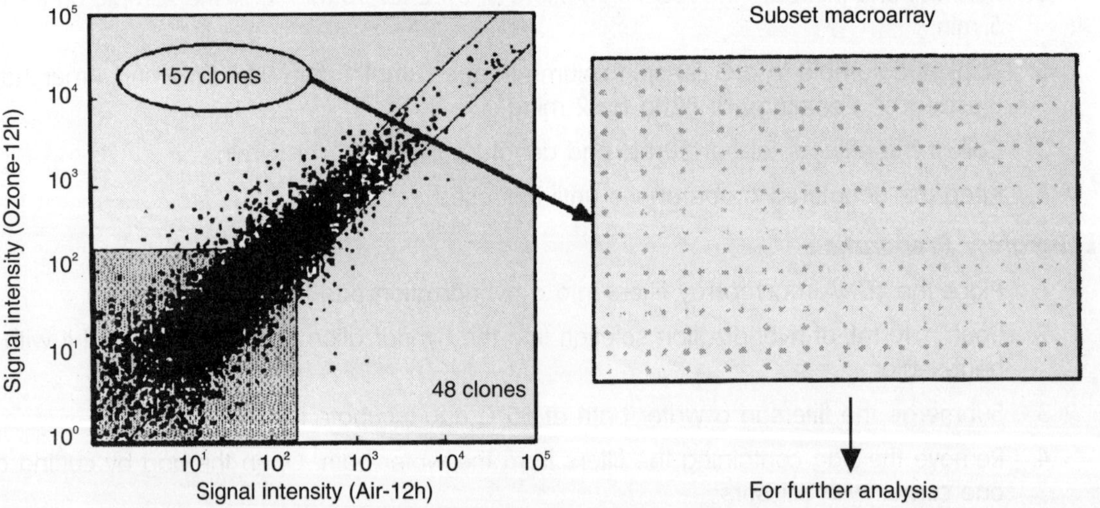

Figure 14.3: Scatterplot of signal intensities of all expressed sequence tags (ESTs) in the Arabidopsis *macroarray* (left) *and the subset macroarray* (right). *Normalized expression ratios for each clone in the macroarray are plotted against signals from 0_3-unexposed (air; x-axis) and -exposed plants (y-axis). The diagonal lines are cutoff lines at 3-fold induction or repression of gene expression. The dots in the shaded area represent ESTs whose signal intensities are not more than 10 times that of the background signal intensity. The 157 clones up-regulated by $_{O3}$ were used to make the subset macroarray.*

stimuli-response genes (O_3-responsive genes in the example experiment), first eliminate from analysis all genes whose expression ratios are less than 10 times the background signal (the average expression ratio of X DNA); this level corresponds to 0.02% expression in total RNA. In the example experiment, this process removed about two-thirds of the 12,028 ESTs from analysis.

5. Select for further analysis genes whose expression was increased or decreased at least 3-fold after treatment stimuli.
6. Genes that respond to stimuli with reproducible results are identified by one-way analysis of variance at a significance level of 0.05. This procedure involves the use of the F statistic, which is used for estimation of population variance on the basis of the information in two or more random samples, to test the statistical significance of the differences among the obtained mRNA expression levels under each experimental condition.

In the example experiment, I identified 205 nonredundant ESTs that were regulated by 0_3. The expression of 157 of these was induced, and that of 48 suppressed, by O_3.

7. (Optional) If hierarchical clustering analysis and K-mean clustering are necessary, the GeneSpring software package (version 5.0, Silicon Genetics, Redwood, CA) is recommended.

8. In the case of *Arabidopsis thaliana,* search for names and annotations of genes by using the Database for *Arabidopsis* Research and Tools (DART) program (http://tabacum.agr.nagoya-u.ac.jp/dart/). Functional classification of genes is performed by first classifying ESTs by referring to MATDB (http://mips. gsf.de/proj/thal/db/index.html). Second, unclassified ESTs are re-sorted according to the putative functions assigned through annotation.

Subset Macroarray Filters

If genes that respond to various stimuli have been isolated, preparation of a subset macroarray using the isolated genes is recommended for further analysis. Because, in general, only about 10% of genes respond to the stimuli of interest, subset macroarrays are more convenient than the large-scale macroarray for further experiments. The subset macroarray typically contains 100-1,000 genes, but I have made a subset macroarray that comprised only 12 genes. The potential applications of subset macroarrays are diverse. For example, subset macroarrays are suitable to confirm whether the isolated genes correctly respond to target stimuli, because investigation of gene expression by using Northern blotting or real-time PCR analysis is time-consuming and costly. Moreover, subset macroarrays are useful to compare the differences in stimulus-associated gene expression patterns among mutant strains or different tissues.

1. Obtain cDNA clones corresponding to genes of interest from a public repos-itory or clone them in-house. In the example experiment, the 157 O_3-upregulated ESTs from *Arabidopsis* were obtained from the Kazusa DNA Research Institute (Kisarazu, Japan).
2. PCR-amplify the inserts of these EST clones by using the primers 5'-GTTTCCCAGTCACGAC-3' and 5'-CAGGAAACAGCTATGAC-3' (these primers are annealed to pBR derived plasmid vectors) and the PCR conditions described elsewhere in this chapter.
3. Mix 75 μL of PCR products with 15 μL of array BPB solution and spot onto a 9 × 12-cm nylon filter (Biodyne A, Pall) in duplicate by using a Multi Pin Blotter 96 (Atto). An example

of subset macroarray filter is shown in Figure elsewhere in this chapter.

4. Spot X DNA (10 mg/µL) in duplicate as a negative control.
5. Fix spotted cDNAs as described earlier.
6. Hybridize the filter and quantify signal intensity as described earlier.
7. Obtain the signal intensity of each spot by subtracting the mean signal intensity of the negative control (X DNA) from the mean signal intensity of the duplicate spots.
8. Normalize the signal intensities of each gene according to that of a gene whose expression level remains unchanged with stimuli treatment. In the example experiment, *AtTub4* was used as a normalization marker gene because its expression level was confirmed previously to be unchanged between stress and nonstress conditions.

POINTS TO REMEMBER

1. All reagents should be prepared using double-distilled water that has a resistance higher than 17.6 Mfl-cm.
2. For preparation of RNAs, glassware must be made RNase-free by baking at 180°C for 8 h or more. All solutions should be made with 0.1% DEPC-treated water, with the exception of the buffer. The primary source of contamination with RNase is the hands of the researcher. Disposable gloves should therefore be worn during manipulations involving RNA.
3. DEPC reacts rapidly with amines and cannot be used to make solutions containing buffers such as Tris. Reserve a fresh, unopened bottle of Tris crystals for preparation of RNase-free solution.
4. DEPC is suspected to be a carcinogen and should be handled with care.
5. The concentration of the RNA can be determined by measuring the OD_{260} of an aliquot of the final preparation. An RNA solution whose OD_{260} is 1 contains approx 45 pg/mL.
6. 5'-methyl dCTP instead of dCTP is used for synthesis of the first-strand cDNA to make the internal *XhoI* sites resistant to digestion.
7. The gene II endonuclease of phage FI (M13 gpII protein; Life Technologies) is no longer commercially available. Instead of this enzyme, site-specific nickase, N.BstT9 (BIORON GmbH, Ludwigshafen, Germany), is available for this reaction. For making a nick in plasmid DNA, mix 5 µg of plasmid DNA, 20 U of N.BstT9 and N. Bst9-buffer [BIORON GmbH, 10 m*M* Tris-HCl, pH 8.5, 10 $_m$*M* $MgC1_2$, 150 m*M* KCl, 1 mM dithiothreitol (DTT), 0.1 mg/mL of bovine serum albumin (BSA)], and the volume adjusted to 50 pL with water. Then incubate the reaction mixture for 30 min at 55°C.
8. The number of clones picked out is dependent on genome size and/or purpose for experiment. If you want to prepare a perfect set of cDNAs that express in *Arabidopsis thaliana* (whole genome size is about 1.25 $\times 10^8$ bp), it should be necessary to pick up more than 3 5 10^6colonies with using normalized cDNA library (average cDNA size is about 2 kb). In general, picking up clones more than 5-fold of genome size is needed to obtain a perfect set of cDNAs.

9. For comparing multiple sets of gene expression data, filters. used in macroarray analysis are often reused. An effective membrane-stripping protocol is boiling the membrane for 30 min in 1 % SDS. This procedure limits reuse to five times because of physical damage to the membrane and reduction in the amount of cDNA spotted on the membrane. A gentler stripping protocol is described by Horngerg et al. Because of the lower melting temperature of the shorter cDNA strands, after oxidative breakdown, stripping by their method can be carried out at relatively low temperatures, thereby reducing heat-induced damage. According to this paper, the same filter is able to use more than 10 times without a measurable reduction in their performance.
10. The hybridization efficiency of macroarray filters can be affected by the strictness of DNA probe, quality of RNA, and number of times the filter has been reused. With high-quality RNA and new filters, the hybridization conditions described elsewhere in this chapter.

Chapter 15

Multiple Alignment of RNAs

RNAs are nucleic acid *polymers* consisting of covalently bound nucleotides. RNA is primarily made up of four different bases: adenine, guanine, cytosine, and uracil. Single-stranded RNA molecules tend to form hydrogen bonds resulting in spatial arrangements of these nucleotides. Many RNAs conserve a secondary structure of *basepairing* interactions more than they conserve their sequence. Since the discovery of RNAs that act as enzymes and the detection of huge classes of *noncoding* RNAs involved in *regulation* processes, RNAs became more and more important.

For the discovery of RNA classes, multiple sequence structure alignments are the best choice to detect RNAs with the same function. Furthermore, multiple *alignments* are an essential prerequisite to further analyses such as *homology modeling,* motif description, or illustration of conserved or variable binding sites.

Here, we want to focus on the concepts and the methods used in multiple alignment of RNAs (MARNA). MARNA is an approach to align multiple RNAs taking into consideration both the primary sequences and the *secondary* structures. It is based on pairwise sequence-structure comparisons of RNAs. From these sequence-structure alignments, *libraries* of weighted alignment edges are generated.

The weights reflect the sequential and structural conservation. For sequences whose secondary structures are missing, the libraries are generated by *sampling* low energy *conformations*. The libraries are then processed by a consistency-based multiple *alignment* method, which is implemented in the T-Coffee system. In addition, MARNA is able to extract a *consensus-sequence* and -structre from a multiple alignment.

Suppose that one has a set of RNA sequences provided with secondary structures. In *summary,* the coarse grain of the MARNA method is as follows:

1. Generate and weight alignment edges between pairwise RNAs reflecting sequence and structure similarities.
2. Collect all weighted edges in a so-called library. This library is processed by a consistency-based multiple alignment method.

3. Find a consensus-sequence and structure from this multiple alignment.

MARNA is able to align RNAs without known conformations as well. For these sequences, several methods to assign structures to the sequences exist. MARNA is *capable* of *integrating* these methods and, thus, to align RNAs with initially unknown *structures*.

In this work, we focus on the methods used in MARNA and give some hints about *parameter* settings that determine the alignments, and structure choices, especially when structures are missing for some sequences.

PROCEDURES

1. A sequence S is a word over the alphabet {A,C,G,U}. S[i] denotes the i-th symbol in S.
2. An arc is a pair $(i, j) \in \{1,..., n\} \times \{1,..., n\}$ s.t. $i < j$. i and j are the ends of the arc. An arc represents a basepair.
3. A base is called free, if it is not involved in any arc.
4. A secondary structure is a set of arcs P, s.t. for any two arcs $(i_1, j_1), (i_2, j_2) \in P$ with $i_1 < i_2$ either $i_1 < j_1 < i_2 < j_2$ or $i_1 < i_2 < j_1 < j_2$.
5. An RNA is a tuple (S, P), where S is the sequence and P is the set of arcs in a secondary structure.
6. An alignment A of two RNAs (S_1, P_1) and (S_2, P_2) is a subset of $\{1, ..., |S_1|\} \cup \{-\} \times \{1 ... , |S_2|\} \cup \{-\}$, where for all pairs $(i, j), (i', j') \in$ A holds

   a. $i \leq i' \Longrightarrow j \leq j'$

   b. $i = i' \neq - \Longrightarrow j = j'$ and

   c. $j = j' \neq - \Longrightarrow i = i'$.

   Requirement: for every $i \in \{1, ... , |S_1|\}$ there is some j with $(i, j) \in$ A (and vice versa for $j \in \{1, ... , |S_2|\}$).
7. The pairs $(i, j) \in$ A are called alignment edges.
8. An alignment edge is called realized if neither $i = -$ nor $j = -$.

Scoring of Alignment

The scoring of an alignment A of two RNAs (S_1, P_1) and (S_2, P_2) is based on the notion of edit operations on bases as well as on arcs. We recall the edit operations and present a slightly modified scoring scheme to finally compute an optimal alignment between two RNAs. Optimal means to find an alignment with minimum costs assuming that the costs of an alignment are composed of the costs of all executed edit operations.

Method of Editing

1. Edit operations on free bases are:
   a. Base match: the base at position i in the first RNA is matched with the base at position j in the second RNA, i.e., $S_1[i] = S_2[j]$. The costs are 0.

b. Base mismatch: the base at position i in the first RNA is aligned with the base at position j in the second RNA s.t. $S_1[i] \neq S_2[j]$. The costs are positive.

c. Base deletion/insertion: the base at position i in the first RNA is aligned with a gap (deletion operation). The opposite case is the insertion operation. Both costs are positive.

2. Edit operations on arcs: consider an arc $(i, j) \in P$, such that i is aligned with i' and j is aligned with j' for $i', j' \in S_2 \cup \{-\}$.
   a. Arc match: an arc match occurs if i', j' form an arc $(i', j') \in P_2$ and $S_1[i] = S_2[i']$ and $S_1[j] = S2[j']$
   b. Arc mismatch: an arc mismatch occurs if i', j' form an arc $(i', j') \in P_2$ and $S_1[i] \neq S_2[i']$ or $S_{1[j]} \neq S_2[j']$
   c. Arc deletion: arc deletion means that $(i', j') \notin P_2$. Depending on how many gaps the two positions i', j' occupy, we may have
      i. Arc breaking: an arc breaking occurs if none of i' and j' equals the symbol –.
      ii. Arc altering: an arc altering occurs if exactly one of i' and j' equals the symbol –
      iii. Arc removing: an arc removing occurs if both i' and j' are equal to –.

3. Edit operation on arcs are as follows:
   a. An arc match has costs 0.
   b. An arc mismatch operation has costs $W_{am}(i, j, i', j')$ for two arcs $(i, j) \in P_1$ and $(i', j') \in P2$.
   c. An arc deletion operation has costs $W_{ad}(i, j, i', j')$. These costs are determined by the bases and by the number of gaps involved. We decompose the costs $W_{ad}(i, j, i', j')$ into a sum of two single functions for the left and right ends of the arcs:

$$W_{ad}(i, j, i', j') = W'_{ad}(i, j) + W'_{ad}(i', j')$$

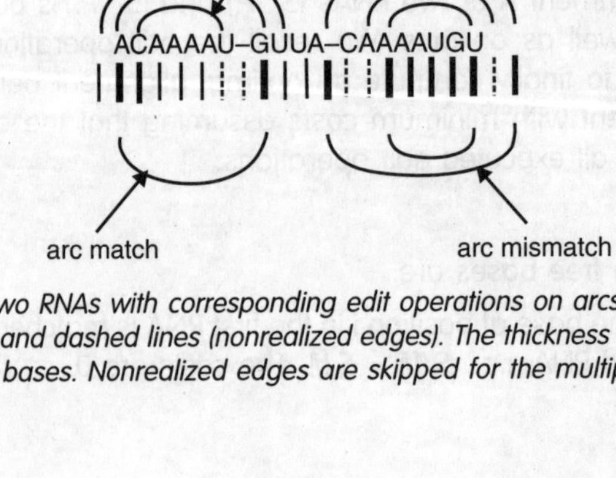

Figure 15.1: An alignment of two RNAs with corresponding edit operations on arcs. Alignment edges are drawn as solid lines (realized edges) and dashed lines (nonrealized edges). The thickness of realized edges corresponds to similarity weights between bases. Nonrealized edges are skipped for the multiple alignment step.

In the following, we do not distinguish between left and right arc ends, and, thus, introduce the function $W^e_{ad}(i, j) = W_{ad}(i, j)+W^r a_d(i, j)$. We even simplify the scoring scheme further by defining $W^e a_d(i, j)$ to be composed of a base match, base *mismatch*, or base deletion together with a fixed cost for deleting an arc. Hence, we set

$$W^e_{ad}(i,j) = W_{base}(i,j) + \frac{1}{2}W^{const}_{ad}.$$

where W_{ad}^{const} are the costs for deleting one arc.

Algorithm

In the following, we specify our *algorithm* that computes an optimal alignment between two RNAs with given secondary structures (S_1, P_1) and (S_2, P_2). We introduce two simple functions:

$$\Psi_S(i) = \begin{cases} 1, \textit{if base at position i not free} \\ 0, \textit{otherwise} \end{cases} \quad (1)$$

$$\chi(i,j) = \begin{cases} 1, \textit{if } S_1[i] \neq S_2[j] \\ 0, \textit{otherwise} \end{cases} \quad (2)$$

Here, the costs for the edit operations on free bases base match, base mismatch, and base deletion are combined into a single cost function $W_{base}(i, j)$, where $W_{base(i,j)} = 0$ only if $S_1[i] = S_2[j]$. Now, we can specify the alignment algorithm:

Input: two RNAs (S_1, P_1) and (S_2, P_2).

Output: sequence structure alignment.

Method: *ALIGN - RNAs()**

for $a_1 = (i_1, i_2) \in P_1$ *and* $a_2 = (j_1, j_2) \in P_2$

dofor $i \leftarrow i_1+1$ *to* i_2-1

dofor $j \leftarrow j_1 + 1$ to $j_2 - 1$

do

$$M(i,j) = \min \begin{cases} M(i-1,j) + w_{base}(i,-) + \Psi_1(i)\frac{1}{2}w^{const}_{ad}, \\ M(i,j-1) + w_{base}(-,j) + \Psi_1(j)\frac{1}{2}w^{const}_{ad}, \\ M(i-1,j-1) + w_{base}(i,j) + (\Psi_1(i) + \Psi_2(j))\frac{1}{2}w^{const}_{ad}, \\ M(i'-1,j'-1) + B(a_k,a_1) + (\chi(i',j')\cdot\chi(i,j))w_{am}(i',i,j',j'), \\ \quad \textit{if } a_k = (i',i) \in P_1 \textit{ and } a_1 = (j',j) \in P_2 \end{cases}$$

$B(a_1, a_2) = M(i_2 - 1, j2 - 1)$

1. We need two two-dimensional *matrices,* both not exceeding the size of *nm*. The matrix *B*

contains the minimum costs of aligning the intervals $(i_1 + 1, i_2 - 1)$ and $(j_1 + 1, j_2 - 1)$ for arcs $a_k = (i_1, i_2) \in P_1$ and $a_1 = (j_1, j_2) \in P_2$ provided that both arcs are aligned; i.e., we have an arc match or arc mismatch. The matrix M *is* constructed when the two arcs a_k and a_1 are considered.

It is computed within the arc intervals in almost the same manner as a sequence alignment except that arc breaking costs are considered and computed at each single base. The algorithm proceeds from inside to outside, thereby taking arcs with minimal sequence lengths first.

2. From the previously described algorithm it is easy to see that the time complexity of $O(n^2m^2)$ results from running over the arcs in both sequences and computing the best alignment in between. The space *complexity* is determined by the sizes of the two matrices B and M.
3. The resulting alignment can be obtained by a traceback step.

Alignment Weights

The alignment algorithm computes an alignment between two RNAs, which is equivalent to an edit transcript composed of edit operations weighted with edit costs. For the multiple alignment step, these costs have to be *transformed* into similarity weights.

1. Note that the costs are *a* function d with positive values fulfilling the metric conditions:
   a. $d(S_1, S_2) = 0 \Leftrightarrow S_1 = S_2$, i.e., the costs of two RNAs S_1 and S_2 are 0 if and only if the two RNAs are equal.

   b. $d(S_1, S_2) = d(S_2, S_1)$, *i.e.*, the edit transcript of transforming S_1 into S_2 has the same costs as the edit transcript of transforming S_2 into S_1.

   c. $d(S_1, S_3) \leq d(S_1, S_2) + d(S_2, S_3)$, i.e., the costs of transforming S_1, into S_2 into S_3 are at least so high as the costs of transforming S_1, into S_3 directly.
2. Transformation from distances to similarities:
   a. *Realized and nonrealized edge*

   Alignment edges are constructed by means of edit operations. Nonrealized edges, denote alignment edges that have exactly one gap at one of their ends. They are skipped for the multiple alignment step because they contain no information about aligning two *nucleotides*. Hence, we are left with realized edges. The thickness *corresponds* to the similarity weights.

   b. *Similarity weights*

   Similarity weights are assigned to edit operations computed by the alignment algorithm. Here, we consider the number of nucleotides r involved in an edit *operation*. We call this number the order of the edit operation. In our case, we have edit operations with

   i. $r = 4$ for an arc match or an arc mismatch.

   ii. $r = 2$ for a base match or a base mismatch.

   iii. $r = 1$ for a base deletion.

Because we have split the arc deletion operation into two separate edit operations for the arc ends, we have an edit operation with r = 2 if the arc end is aligned with a *nucleotide*, and an edit operation with $r = 1$ if the arc end is *aligned* with – The similarity weights can be achieved by choosing a maximal similarity value *M*, such that every value can be subtracted from the value $r \cdot M$. The value *M* is multiplied by *r* because we, therefore, ensure that all similarity values are positive.

Multiple Alignment

Now, we are ready for the multiple alignment step. Suppose we have set of *n* RNAs together with their secondary structures. The main idea is to use the same strategy as proposed by the multiple alignment tool T-Coffee.

1. Recall that a single alignment between two RNAs provides a set of weighted, realized alignment edges.
2. The pairwise comparison strategy in a set of *n* RNAs yields $n(n-1)/2$ alignments. All these alignments produce an amount of weighted alignment edges each reflecting the sequence structure similarity between two bases. These edges are collected in a so-called library.
3. Now, the T-Coffee strategy is performed on this data set:

    a. *Library extension:*

    The library containing all pairwise alignments with their weighted alignment edges is turned into an extended library to improve all pairwise alignments by taking into *considerations* how all other sequences align with the current two. For instance, if we consider two RNAs specified by their alignment and their weighted *alignment* edges then a third sequence is considered how this sequence is aligned with the first and the second sequence.

    For any alignment of two RNAs R_1 and R_2, any other RNA R_3 is considered for improving the initial alignment. For this purpose, T-Coffee considers the alignment of R_1 and R_2 via R_3 by considering alignment edges from the alignment of R_2 and R_3 with edges from the alignment of R_2 and R_3. These additional weighted edges together with the edges of the direct comparison of the first two RNAs are considered to improve this alignment by a dynamic programming approach. This procedure is executed $n(n-1)/2$ times, i.e., for each pairwise set of RNAs. The result is the *extended* library containing all improved pairwise alignments.

    b. *Progressive alignment:*

    Pairwise distances of the sequence set were computed owing to the alignment algorithm. They form the distance matrix which is used to produce a *neighbour*-joining tree that guides the alignment process. Residue weights that are stored in the extended library are now used for this task. The two closest sequences are aligned first.

    This alignment is fixed and the next closest sequence is aligned to this existing alignment or two new sequences are aligned or two existing alignments are aligned. In the case of aligning an already existing alignment, the average score in each

Edit-Op	Name	Distance	Similarity
A···U A···U	arc match	0	4 .M
A···U G···C	arc mismatch	$w_{am}(A,U,G,C)_A$	$4 .M - w_{am}(A,U,G,C)$
A··· G···	arc breaking arc altering (realized edge)	$w_{base}(A,G,) + \frac{1}{2} w_{ad}^{const}$	$2 .M - w_{base}(A,G) - \frac{1}{2} w_{ad}^{const}$
A··· G···	arc breaking arc altering (realized edge, two arcs)	$w_{base}(A,G,) + w_{ad}^{const}$	$2. M - w_{base} (A,G) - w_{ad}^{const}$
···	arc breaking arc removing (non–relaized edge)	$w_{base}(A,–) + \frac{1}{2} w_{ad}^{const}$	no realized edge

Figure 15.2: Edit operations.

column is taken. We do not need gap penalties because they are already included in the alignment as sequence identities and residue weights, i.e., residues which are aligned with gaps get a weight of zero.

Experimental Approach

The multiple alignment of these RNAs assumes the existence of a known structure for each RNA like, e.g., an experimentally confirmed structure.

1. Whenever the structures are not known in advance, secondary structure prediction programs like Mfold and RNAfoldA may help to assign the minimum free structure to an RN The *drawback* here is that these structures are not necessarily the real existent structures which might be responsible for their *functions*.
2. To overcome this difficulty, we assign multiple structures to each sequence covering different folds. We call this set the *ensemble* of *structures*. We mainly use two different programs for generating these structure:

a. RNAsubopt : this program generates *suboptimal* structures by *stochastic backtracking*. The number of desired structures can be set *individually*.

b. RNAshapes : this program avoids the large output of similar suboptimal structures; instead, it outputs structures of more fundamental differences.

3. The generated structures for each sequence S_l form an ensemble, denoted $E^l_S = \{E^l_1, \ldots, E^l_r$ Because each structure E^l_i has its own energy, it occurs with probability, say $\Pr(E^l_i)$. Here, we consider rather a small set of important structures, in contrast to the explosive number of all suboptimal structures.

4. Probability: because of their different energies of these structures assigned to sequence S_l, the probability of seeing a certain structure E^l_k in a set of structures E_{Sl} with restricted size n is:

$$\Pr\left(E_k^l \middle| E_{S_l}\right) = \frac{\Pr\left(E_k^l\right)}{\Sigma_{i \le i \le n} \Pr\left(E_i^l\right)} \tag{3}$$

where $\Pr(E^l_k)$ is the probability of forming structure E^l_k in sequence S_l. In MARNA, the simplification of the uniform distribution is made, i.e., each structure has the same *probability*.

5. Alignment weights: consider two sequences S_1 and S_2 with *n*, structures for the first sequence and n_2 structures for the second sequence. If both $n_1 = 1$ and $n_2 = 1$, then the alignment algorithm outputs weighted alignment edges as we have seen before. These *alignment* edges are all *multiplied* by one because the number of structures in the ensemble equals one.

Suppose we consider an ensemble of structure greater than one, then we have to make $n_1 \times n_2$, comparisons, i.e., each combination of (S_1, E^1_k) and (S_2, E^2_l), $1 \le k \le n_1$, $1 \le l \le n_2$, has to be considered. The number of realized alignment edges is quadratic, i.e., proportional to $n_1 \times n_2$. The alignment weights are now influenced by the structural diversity. Each alignment edge is reweighted by the factor $\Pr(E_k|E_{sl})\ Pr(E^2_l|E_{s2})$. If both $|E_{s1}| = 1$ and $|E_{S2}| = 1$, then the alignment edges are weighted by the factor 1.

Final Alignment

Once we have computed the final alignment, we are ready to calculate a consensus structure from this alignment. Here, we explicitly use structure information for the calculation of the alignment. Hence, the calculation of the consensus structure should be based on these ensemble structures.

1. To *exemplify* the basic idea, suppose that exactly one structure per sequence is given. Each structure must then be *interpreted* as the *"real"* known structure. A conserved basepair between two columns in the alignment is found if the majority of sequences have a basepair at the corresponding *sequence positions*.

   The remaining problem is that the resulting set of conserved *basepairs* alone does not form a secondary structure and is thus not a valid consensus structure. This is a problem common to all approaches for calculating a *consensus* structure.

2. We find a remedy by calculating a consensus secondary structure that maximizes basepair conservation. So let c, c' be two columns with $1 \chi\ c \leq c' \leq m$, where m is the number of columns of the multiple alignment. Furthermore, let $bp\_cons(c, c')$ be the number of sequences that have a basepair between the corresponding sequence positions. The consensus structure is then defined to be a secondary structure $P \subseteq [1..m] \times [1..m]$ such that

$$\sum_{(c,c')} bp\_cons(c,c')$$

is maximized.

3. This can be calculated using *dynamic* programming. Let $N_{i,j}$ with $1 \leq i, j \leq m$ be the maximal *basepair* conservation for all columns between i and j:

$$N_{i,j} = \max_{P} \sum_{\substack{(c,c')\in P \\ i\leq c\leq c'\leq j}} bp\_cons(c,c')$$

The corresponding recursion equation for $N_{i,j}$ is

$$N_{i,j} = \max \begin{cases} N_{i+1,j} \\ N_{i,j-1} \\ N_{i+1,j-1} + bp\_cons(i,j) \\ \max\limits_{i<k<j} \{ N_{i,j+k} + N_{i+k+1,j} \} \end{cases}$$

It is a *dynamic* programing *approach*, where the traceback reports the consensus structure of the alignment.

4. Finally, we have to consider again the case where we are given structure ensembles for some (or all) sequences. Consider a multiple alignment of K sequences. For each sequence S_k, let E_{sk} be the ensemble of structures calculated for S_k. For each column c, let i^k_c be either the position that corresponds to column c in sequence S_k (if aligned), or - otherwise. Furthermore, let $\delta_P(c, c')$ *be* the index function of *P, i.e.,* $\delta_p(c, c')$ is one if $(c, c') \in P$, and 0 otherwise. Then

$$bp\_cons(c,c') = \sum_{k=1}^{K} \sum_{E^k_j \in E_k} \delta_P\left(i^k_c, i^k_{c'}\right) \Pr\left(P \middle| E_{S_k}\right)$$

where $\Pr(P|E_{sk})$ is defined as given in Eq. 3.

POINTS TO REMEMBER

1. MARNA can be tested online via the webpage http://www.bioinf.unifreiburg.de/Software/MARNA/index.html. MARNA is also available as a downloadable file.
2. MARNA offers mainly two choices to adjust the alignments:
   a. *Parameter settings*: MARNA relies on the comparison of pairwise RNAs. These comparisons are accomplished by alignments with costs assigned to edit operations on bases and arcs. These costs can be set individually.

   b. *Structure computation:* The alignment of RNAs take into account both the primary sequences and the secondary structures. The easiest case is when the secondary structures are known in advance, and the computation is reduced to find common *sequential* and structural properties.

   Otherwise, the structures have to be found. MARNA provides in addition to user-defined structures the assignment of different sets of structures. These include the assignment of minimum free energy structures, shaped structures or an ensemble of low energy structures.

3. *Parameter setting:*

   Parameters can be set individually depending on weighting some edit operations more or less. A series of tests has brought three data sets to obtain alignments based on sequential or structural properties or on a mixture on both.

4. Parameters settings influence the resulting alignments. Choose the default parameter settings first. It has been confirmed that this data set recognizes conserved sequential and structural properties very well.

5. Beyond the parameter settings, the assignment of different structures to the sequences are quite important as well. The easiest case is when user-defined structures are given as input.

6. Structure choice: here are some hints to choose the right structure assignments if no structures are given to the sequences.

   a. If the RNAs are sequentially related and have nearly the same length then choose the minimum free energy structures.

   b. The shaped structures are suited to cover a lot of diverse structural conformations for each single sequence. Choose shape structures, if no clear consensus structure is observable at first glance.

   c. The ensemble set of low energy conformations is the best choice if it is assumed that these RNA sequences resemble structurally in some way. An ensemble consists of multiple structures. This ensemble contains similar structures if almost all suboptimal structures are similar.

7. The running time of MARNA crucially depends on the structure choices. Suppose n RNAs of nearly the same length without structure specifications are given. If the mfe structures are chosen that are assigned to the sequences then the multiple alignment and the consensus structure computation can be done in reasonable time.

   Suppose the user chooses an ensemble of three suboptimal structures to each RNA, then the computation time is ninefold because for each pair of RNAs nine pairwise sequence structure comparisons have to be made.

Chapter 16

Bead Arrays Technology

Illumina's BeadArray technology is based on the random *self-assembly* of a bead pool onto a patterned substrate. The arrays are assembled into a matrix (the Sentrix" Array Matrix, 96 samples, or Sentrix BeadChip. 16 samples), allowing efficient processing of thousands of samples.

The manufacture of *Illumina microarrays* has been described elsewhere. Briefly, universal BeadArrays are assembled by loading pools of glass beads (3 μm in diameter) derivatized with oligonucleotide capture probes onto the etched ends of fiber-optic bundles or the etched surface of silica slides. Because the beads are positioned randomly, a *decoding* process is carried out to determine the location and identity of each bead in every array location. Decoding is an automated part of array *manufacturing* and has the added benefit of *validating* the *hybridization* capability of each feature.

Gene Expression Profiling

There were over 300 million archived tissue samples were in the United States in 1999, with more samples accumulating at a rate of over 20 million per year. These samples represent an extraordinary resource for basic research as well as for biomarker discovery and validation, yet they are largely inaccessible to standard microarray technology due to extensive degradation of the RNA. In contrast to most array *technologies* that use an in vitro transcription (IVT)-*mediated* sample labeling procedure, the DASL Assay uses random priming in the cDNA synthesis and, therefore, does not depend on an intact poly-A tail for oligo-d(T) *priming*.

The DASL Assay interrogates specific cDNA sequences with *oligos* linked to unique address sequences that can hybridize to their *complementary* strand on the universal arrays [10]. The assay requires a relatively short target sequence of about 50 *nucleotides* for query oligonucleotide annealing.

Thus, the DASL Assay allows analyses of partially degraded RNAs extracted from FFPE tissues. The DASL Assay uses one set of universal PCR primers to amplify all the targets and generates amplicons of ~100bp. This *uniformity* results in a relatively unbiased amplification of the PCR template *population*. The approach combines the advantages of array-based gene expression analysis with those of *multiplexed* qPCR, thereby *offering* much higher *multiplexing* capacity and

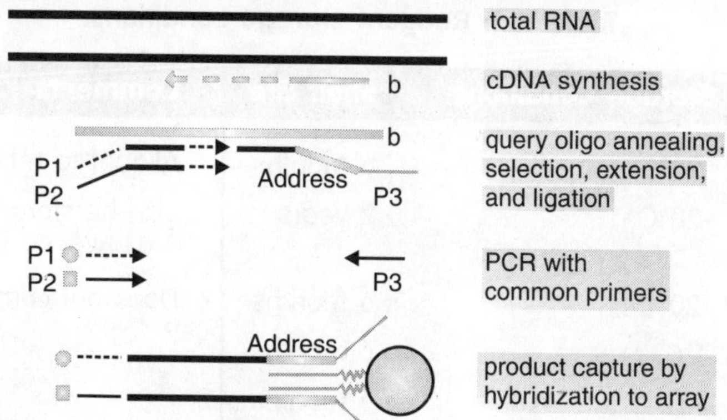

Figure 16.1: DASL Assay scheme. In the DASL Assay, three oligos are designed to target a specific gene sequence. The total RNA first is converted to cDNA by random priming. The corresponding query oligos bind to the cDNA and become extended and ligated enzymatically. The ligated prod-ucts then are amplified and fluorescently labeled during PCR and finally detected by hybridization to address sequences on the array.

throughput at lower cost. The assay multiplexes to over 1,500 sequence *targets*, and expression differences obtained with this assay compare well to those *determined* by qPCR.

Cell-specific expression was readily detected, and expression data allowed biologically appropriate sample classification. With automation, the DASL Assay coupled with the array readout can be used for high-throughput expression profiling of hundreds of genes in hundreds to thousands of FFPE samples. This opens up new avenues to large-scale *discovery, validation*, and *clinical application* for expression markers of disease.

REQUIREMENTS

The Materials and Methods sections assume that the user has access to the Illumina BeadStation or BeadLab systems and associated reagents and equipment.

Kit for Extraction of RNA

Of the few RNA preparation kits we tested, the High Pure RNA Paraffin Kit from Roche Applied Science yielded the highest quality RNA from FFPE samples for use in the DASL Assay. Recently, more RNA *extraction* kits have become available. If a different kit is used for RNA extraction, the *prequalification* steps described in the Notes section should be used to increase the *likelihood* of success in the DASL Assay.

Reagents Required

1. MCS/MCM (master mix for cDNA synthesis for single or multiple use), reagent for cDNA synthesis.
2. DAP (DASL Assay oligo pool), mixture of *oligonucleotides* designed to query cDNA target sequences.
3. OB I (oligo binding buffer 1), oligo *annealing* buffer that also contains *paramagnetic* particles to optimize the washing, extension, and ligation steps of the assay.

Table 16.1: Reagent storage conditions.

Reagent	Storage Conditions	Shelf Life	Comments
MCSIMCM	-20°C	6 months	Aliquot to refreeze
DAP	-20°C	2 years	Can he stored at 4°C up to 2 weeks
OB1	-20°C	6 months	Does not completely freeze
AM 1	4°C	6 months	-
UB 1	4°C	6 months	
MEL	-20°C	6 months	Aliquot to refreeze
MMP	-20°C	6 months	Aliquot to refreeze after adding DNA polymerase
IP1	-20°C	1 year	
MPB	4°C	6 months	Do not freeze
UB2	RT	6 months	
MH1	RT	6 months	Keep away from light
ISI	Room temperature while dry -20°C when reconstituted	6 months	Must resuspend at least 1 d in advancem
WC1	-20°C	6 months	Must dilute
AC 1	Room temperature	6 months	

4. AM 1 (add MEL buffer 1), wash buffer used to *remove* excess of mis- or unhy *bridized* query *oligonucleotides.*
5. UB 1 (universal wash buffer 1), wash buffer for several pre-PCR steps.
6. MEL (master mix for extension and ligation), optimized mixture of enzymes for the *extension* and *ligation* steps.
7. MMP (Master mix for PCR), PCR master mix that contains *fluorescent* common *primers* for the *multiplexed* ligated oligo templates.
8. IP1 (inoculate PCR buffer 1), elution buffer for inoculating the PCR reaction with the ligated templates.
9. MPB (magnetic particle buffer B), a suspension of paramagnetic particles used to bind PCR products.
10. UB2 (universal buffer 2), a wash buffer used in post-PCR processes.
11. MH 1(make Hyb buffer 1), a neutralization buffer used in making the final sample single-stranded for hybridization to the Sentrix Array Matrix or BeadChip.

12. IS1(image SAM buffer 1), a buffer needed to dry bundles and protect fluors before imaging.
13. WC1 (wash chip buffer 1), a buffer used to wash Sentrix BeadChips after hybridization.
14. AC 1 (add coverslip buffer 1), a buffer used for protecting fluors in imaging the BeadChip.

Please refer to Table elsewhere in this chapter for reagent storage conditions.

Additional Reagents Required

1. 0.1 N NaOH.
2. TE buffer: 10 mM Tris-HCl, pH 8.0, 1 mM EDTA.
3. DNA polymerase (Clontech BD, 639208 or 639209).
4. Uracil DNA glycosylase, l U/μL (Invitrogen, 18054-015).
5. 95% EtOH (v/v).
6. 2-butanol 99% (Sigma-Aldrich, B85919).
7. RiboGreen RNA Quantitation Kit (Molecular Probes Invitrogen, R-11490).
8. Xylenes (EMD Chemicals, XX0055-13) for FFPE sample deparaffinization.
9. Serological pipettes (10, 25, and 50 mL).
10. 96-well 0.2-mL skirted microplates (Bio-Rad, MSP-9601).
11. 0.8-mL storage plate, conical well bottom (ABgene, AB-0765).
12. 96-well, black, flat-bottom Fluotrac 200 plates (Greiner, 655076).
13. 96-well sealing mats, round cap, autoclavable (ABgene, AB-0674).
14. 96-well cap mats, sealing mats, round cap, piercable, nonautoclavable (ABgene, AB-0566).
15. Heat sealing foil sheets, thermo seal (ABgene, AB-0559).
16. Microplate clear adhesive film, 2-mil Sealplate adhesive film, nonsterile (Phenix Research Products, LMT-SEAL-EX).
17. Microseal A film, PCR plate sealing film (Bio-Rad, MSA-5001).
18. Microseal F film, aluminum adhesive film (Bio-Rad, MSF-1001).
19. 96-well V-bottom plates, Corning Costar polypropylene (Fisher Scientific 07-200-695 or VWR, 29444-102).
20. Multiscreen filter plates, 0.45-μm, clear, styrene (Millipore, MAHV-N45 10/50).
21. Nonsterile solution basins (Labcor Products Inc., 730-001 or VWR, 21007-970).
22. Cliniplate 384-well microplates (Thermo Labsystems 95040000).
23. Omni trays (Nunc, 242811 *or* VWR, 62409-600).

Array Matrix and Bead Chip

Illumina produces microarrays in two formats: the Sentrix Array Matrix (SAM) and the Sentrix BeadChip. In the array matrix, the *fiberoptic bundles*, each containing -50,000 5-μm fibers, are arranged into an Array of Arrays format that is compatible with and can access the wells of a 96-well *microtiter* plate. The *fiberoptic* bundles in the array matrix assembly are polished flat on both ends.

On one end, the cores of each fiber are etched to form *nanowells* that will accept 3-μm silica beads. Each of the beads has been derivatized with several hundred thousand oligonucleotides of a particular sequence. In the BeadChip format, several microarrays are arranged on silicon slides that have been processed by *microelectromechanical* systems (MEMS) technology to also have nanowells that support self-assembly of beads.

In both formats, an average of over 30 beads of each type is maintained, a *strategy* that provides the accuracy of multiple measurements and statistical power. Beads are self-assembled in the wells during the *manufacturing* process, and each array is decoded to determine the map of bead type positions.

Combinatorial series of brief *hybridizations* and rinses are used for decoding that result in a level of accuracy well beyond the requirements of any application [3]. This decoding process also provides a quality control measure of the function of each bead that is *incorporated* into the final bead map. BeadArray technology is the only microarray technology with functional quality control of every element in the array.

Array Imaging

The Illumina BeadArray Reader is required for array imaging. Each assay results in a fluorescent signal associated with individual bead types on the array. To read out these signals, we developed the BeadArray Reader, a two-channel, 0.8-μm resolution confocal laser scanner that can simultaneously (via menu-driven software) scan a BeadChip or SAM at two wavelengths and create an image file for each channel. For Illumina BeadArray Reader instructions, see the *Illumina BeadArray Reader User Guide.*

PROCEDURES

RNA Extraction from FFPE Tissues

RNA extraction from formalin-fixed, paraffin-embedded samples is a critical step for success in the DASL Assay. We recommend the High Pure RNA Paraffin Kit (Roche) for RNA purification, using five 5-μm sections of FFPE tissues. All *buffers* and reagents are provided with the kit. The RNA extraction is carried out following manufacturers' *recommendations* (see notes and tables elsewhere in this chapter for details for RNA yields from different tissue sources). RNA integrity can be examined with the Agilent 2100 *Bioanalyzer* using RNA 6000 Pico Assay Kit.

DASL Assay Protocols

These protocols are not intended to replace the DASL Assay *manual* supplied with Illumina systems but rather give a detailed overview of the process.

The DASL Assay is based on the sequence-specific extension and ligation of correctly hybridized query oligos, which are distinguished by their shared primer landing sites (for an overview of the DASL Assay workflow, see http://www.illumina.com/downloads/DASLTECHBULLETIN.pdf). Query oligos are designed to target *nonpolymorphic* sites in cDNAs, and the assay uses two upstream

Table 16.2: RNA yield from five 5-μm sections of FFPE tissue.

Tissue	Diagnosis	Yield (μg)
Prostate	Normal	3.81
Prostate	Neoplasia	2.09
Prostate	Adenocarcinoma	1.31
Lung	Normal	0.34
Lung	Adenocarcinoma	11.45
Lung	Bronchio-alveolar adenocarcinoma	7.80
Lung	Small cell adenocarcinoma	4.73
Lung	Squamous cell adenocarcinoma	13.22
Breast	Normal	0.65
Breast	Fibroadenoma	4.49
Breast	Invasive ductal adenocarcinoma	17.14
Breast	Lobular adenocarcinoma	5.56
Colon	Normal	10.07
Colon	Adenocarcinoma mucoid	2.62
Colon	Adenocarcinoma sygmoid	6.40
Colon	Adenocarcinoma ulcer	14.46

oligos (ASOs) for each downstream oligo (LSO). The result is two *separate* measurements of the population of a single *transcript,* one in each fluorescent channel. Used only for expression *profiling,* the DASL Assay need not be *performed* with two fluors.

However, its current configuration allows the assay to be compatible with the reagents used in the Illumina GoldenGate* *Genotyping Assay.* This arrangement gives the researcher the flexibility to investigate the *allele-specific* expression by "*genotyping*" cDNA as well as genornic DNA. For array analysis, three query oligonucleotides are designed to *interrogate* each target site on the cDNA, with up to ten target sites per gene.

Query oligos are designed to target *intraexonic* sequences, using bioinformatic and biochemical *parameters* optimized for the assay biochemistry. The ASO oligos consist of two parts: the gene-specific sequence and the *universal* PCR primer sequences, with P1 and P2 at the 5'-end.

The downstream oligo consists of three parts: the *gene-specific* sequence, a unique address sequence that is *complementary* to 1 of 1,536 capture sequences on the array, and a universal PCR primer sequence (P3) at the 3'-end.

A single address sequence is uniquely associated with a single target site. This address sequence allows the PCR-amplified products to *hybridize* to a universal microarray bearing the complementary probe sequences.

Laboratory Synthesis of cDNA

1. Preheat the heat block to 42°C and allow temperature to stabilize. Thaw the cDNA *synthesis* reagent (MCS, Illumina) to room temperature and *vortex* to fully mix tube contents.
2. Normalize the RNA samples to 40-100ng/μL with DEPC-treated H_2O. We recommend using 200ng of RNA extracted from FFPE samples for one assay.
3. Add 5μL of MCS to each well of *microtiter plate* labeled SUR. Transfer 5μL of normalized RNA sample to each well of the SUR plate, heat seal with *foil*, and vortex at 2,300 rpm for 20s. *Pulse-centrifuge* the sealed plate to 250g to prevent the wells from evaporating during the *incubation*.
4. Incubate the SUR plate at room temperature at least 10 min, 1 then continue incubation at 42°C for 60 min in the preheated heat block.
5. When incubation is complete, remove the plate from the heat block and pulsecentrifuge to 250g to collect any condensation.
6. If the experiment will be continued the same day, set the heat block to 70°C. If not proceeding with the experiment immediately, the SUR plate can be stored up to 4 h at 4°C or overnight at -20°C.

Synthesis of cDNA from Oligonucleotides

1. Preheat heat block to 70°C and allow temperature to stabilize.
2. Remove DASL Assay *oligonucleotides* (DAP) tube from refrigerator (if frozen, thaw, vortex, then centrifuge). Thaw the oligonucleotide annealing reagent (OB1) to room temperature and vortex. Do not *centrifuge* the OB1 tube.
3. Dispense 10μL of DAP to each well of a new, 96-well, 0.2-mL skirted microplate labeled ASE. Add 30μL of well-resuspended OB1 to each well of that plate.
4. Centrifuge the SUR plate to 250g to collect samples at the bottom of the wells. Transfer l0 gL of the *biotinylated* cDNA to the ASE plate containing the DAP and OB1 to bring the final volume to 50μL. Heat seal the plate and vortex briefly at 1,600 rpm to mix the content of the wells. Place the ASE plate in the 70°C heat block and immediately reduce the temperature setting to 30°C. This will carry out assay oligonucleotide annealing to the cDNA target by ramping the temperature over approximately 2 h then holding at 30°C until the next processing step.

Elongation of Oligonucleotide Chain

1. Remove the ASE plate from the heat block, reset it to 45°C and allow the temperature to stabilize. Thaw the extension and ligation reagent (MEL) to room temperature.
2. Place the ASE plate with assay oligonucleotides annealed to the cDNA template on the Illumina-supplied magnetic plate for at least 2 min, or until the beads are completely captured. Washing the beads removes excess and mishybridized oligonucleotides.
3. After the paramagnetic particles are captured, remove the heat seal from the plate and remove and discard all liquid (~50μL) from the wells, retaining the beads. Add 50μL Add

MEL buffer (AM 1) to each well of the assay plate. Seal the plate with adhesive film and vortex at 1,600 rpm for 20 s or until all beads are resuspended.

4. Place the ASE plate on the magnet for at least 2min, or until the beads are completely captured. Remove all the AMI from each well, leaving the beads in the wells. Repeat the addition of 50 µL AM I, vortexing, and removal of buffer.
5. Remove the ASE plate from the magnet and add 50 gL of universal wash buffer (UB 1) to each well.
6. Place the ASE plate onto the *magnet* for at least 2 min, or until the beads are completely *captured*. Remove all UB 1 from each well. Repeat the addition of 50 µL UB1 and removal of buffer.
7. Add 37 µL of extension and *ligation* master mix (MEL; Illumina) to each well of the ASE plate. Seal plate with *adhesive* film and *vortex* at 1,600 rpm for 1 min.
8. Incubate the ASE plate on the *preheated* 45°C heat block for 15 min.

Method for Polymerase Chain Reaction

1. Prepare PCR master mix by adding 64 µL of DNA polymerase and 50 µL UDG to the tube of PCR reagent (MMP). Invert the MMP tube several times to mix contents and *aliquot* 30µL of mixture into each well of a new 96-well 0.2-mL microplate (the PCR plate).
2. Remove the ASE plate from the heat block after the extension and ligation step and reset the heat block to 95°C.
3. Place the ASE plate on the magnet for at least 2 min, or until the beads are captured. Remove the clear adhesive film from the assay plate, and remove and discard *supernatant* (50µL) from all wells of the ASE plate, leaving the beads in wells. Leave the ASE plate on the magnet and add 50pL of the *universal* wash buffer UB1 to each well of the plate.
4. Allow the ASE plate to rest on *magnetic* plate for at least 2 min to collect the paramagnetic particles. Remove and discard all supernatant (~50µL) from all wells of the ASE plate, leaving beads in wells.
5. Add 35µLof the elution buffer IP1 to each well of the assay plate and seal it with adhesive film. Vortex the plate at 1,800 rpm for 1 min, or until all the beads are *resuspended*. Place the plate on the 95°C heat block for 1 min.
6. Remove the ASE plate from heat block and place it onto the magnet for at least 2 min, or until the beads have been completely captured. Transfer 30µL of supernatant from the first column of the ASE plate into the first column of the PCR plate. Repeat the transfer for the remaining columns, using new pipette tips for each column.
7. Seal the PCR plate with Microseal A PCR plate sealing film. Immediately transfer the plate to the thermal cycler and run the following cycling program: 10 min at 37°C; 34 cycles (35 s at 95°, 35 s at 56°, 2 min at 72°); 10 min at 72°; and 4'C for 5 min.
8. Proceed immediately to the preparation of the PCR product for array hybridization or seal and store the PCR plate at -20°C.

Formation of Array Hybridization

1. Vortex the tube with the suspension of paramagnetic particles (MPB) until the beads are completely resuspended.
2. Dispense 20µL of resuspended MPB into each well of the PCR plate. Mix the beads with PCR product by *pipetting* up and down, then transfer the mixed solution to the *filter plate.* Cover the filter plate with its cover and store at room temperature, protected from light, for 60 min.
3. Place the filter plate containing the bound PCR products onto a new 96-well V-bottom waste plate using a filter plate adapter. *Centrifuge* at 1,000g for 5 min at 25°C.
4. Remove the filter plate lid. Add 50µL of universal wash buffer 2 (UB2) to each well of the filter plate. *Dispense* slowly, so as not to disturb the beads. Relid the filter plate and centrifuge at 1,000 g for 5 min at 25°C.
5. Prepare a new 96-well V-bottom plate and dispense 30 µL of hybridization buffer (MH1) to all wells of the new intermediate plate. Place the filter plate onto the intermediate plate such that column Al of the filter plate matches column Al of the intermediate plate.
6. Dispense 30 µL of 0.1 *N* NaOH to all the wells of the filter plate. Relid the filter plate and centrifuge immediately at 1,000g for 5 min at 25°C. Gently mix the contents of the intermediate plate by moving it from side to side, without *splashing.*
7. Prepare a HYB Cliniplate 384-well microplate. First, dispense 30µL of UB2 buffer in every other well (B2, B4, D2, D4, etc.) of the hybridization plate for humidity control. Then transfer 50µL of *neutralized* hybridization solution from the *intermediate* plate into appropriate wells (Al, A3, C l, C3, etc.) of hybridization plate. It is helpful to use separate templates for *humidity* control wells and hybridization wells to assist in sample *dispensing.*
8. Seal the plate with clear adhesive film and centrifuge at 3,000g for 4min at 25°C. If the hybridization is not performed immediately, the HYB plate can be stored at -20°C.

Converstion of Hybridization to Bead Arrays

1. Preheat the oven to 60°C and allow the temperature to equilibrate. If the HYB plate has been frozen, thaw it *completely* at room temperature in a *lightprotected* drawer. *Centrifuge* the plate at 3,000g for 4min.
2. Prepare two OmniTrays, one with 70 mL of UB2 and another with 60 mL of 0. 1N NaOH. Carefully place the Sentrix Array Matrix (SAM), with the *barcode* facing up and the fiber bundles facing down, into the UB2 tray. Agitate the SAM gently (l0s only) to remove *bubbles* from the *bottoms* of the arrays. After 3 min, move the SAM into a NaOH tray and incubate 30 s, then move the SAM back into the UB2 *conditioning* tray. Allow the SAM to sit in the UB2 conditioning tray for at least 30 s to neutralize the NaOH.
3. Remove the clear adhesive film from the HYB plate. Insert the SAM fiber-optic bundles into appropriate wells of the HYB plate using the SAM alignment fixture or a SAM hybridization *cartridge.* Place the HYB plate/SAM pair into the hybridization oven preheated to 60°C and incubate for 30 min.

4. Hybridization then is conducted under a temperature gradient program from 60 to 45 °C over approximately 12 h.

Technique for Imagning

1. Reconstitute the array drying reagent (IS1; Illumina) by adding 94 mL of an equal mixture of 95% *ethanol* and 2-butanol to 6 mL of IS1 to *prepare* enough reagent to image one array.
2. For each SAM to be imaged, prepare two OmniTrays with 70mL of UB2 and one with 70 mL of diluted IS1.
3. Carefully separate the SAM from the HYB plate. Place the SAM into the first UB2 tray and agitate 1 min at room temperature. Transfer the SAM to the second UB2 tray and repeat the agitation for 1 min at room temperature. Ensure no *bubbles* are on the bottom of the array fiber bundles. These bubbles may prevent certain areas of the *fiber* bundle from contacting the liquid.
4. Dip the SAM into the IS1 tray. It is important to dip the SAM several times to ensure that the UB2 buffer is completely exchanged. After 5 min, remove the SAM from the IS1 and place it on an empty *OmniTray* to air dry 20 min, fiber *bundles* up. The dry array is ready for *imaging*.
5. The arrays are imaged using the BeadArray Reader (Illumina). Image processing and intensity data *extraction* is performed by the BeadScan software included with the BeadArray Reader.

Analysing Data

The BeadStudio software package is included with the Illumina DASL Assay product and is used as a tool for analyzing gene expression data from scanned microarray images collected from the Illumina BeadArray Reader. Alternatively, *BeadStudio* can be used to export the array intensity data for processing by most standard gene expression *analysis* programs.

Specifically, BeadStudio executes two types of data analysis:

1. *Gene analysis*, determining gene expression levels and whether gene *expression* is detected.
2. Differential analysis, determining if gene expression levels are different between any *experimental* groups.

Analysis can be performed on individual arrays or groups of arrays treated as replicates. BeadStudio reports experiment performance based on built-in controls that accompany each experiment. In addition, BeadStudio provides scatterplotting and *dendrogram tools*, facilitating quick, visual means for *exploratory analysis*.

Applications

Our data showed that the DASL Assay was capable of validating tissue and cancer markers previously identified using RNA prepared from fresh-frozen tissues [11]. We also demonstrated that, using the DASL Assay system, highly reproducible tissue- and cancer-specific gene expression profiles could be obtained with RNA isolated from *formalin-fixed* tissues that had been stored from 1 to over 10 years. Furthermore, the marker sensitivity and *specificity* measured by the DASL analysis were comparable to those determined for qPCR. Overall, the DASL Assay analysis *outperformed* qPCR for these samples.

In addition, we constructed a *splicing* array to interrogate > 1,500 mRNA isoforms from a panel of genes previously *implicated* in prostate cancer. We derived a highly specific panel of mRNA isoforms for *prostate* cancer using independent sets of samples for marker selection and validation.

POINTS TO REMEMBER

1. RNA from FFPE specimens can be difficult to extract, since the RNA becomes cross-linked and degraded during the fixation and storage process; in addition, the amount of tissue in the FFPE specimen can be very small (for example, normal lung samples consist mostly of *extracellular* air space). Therefore, it is essential to have a robust method to retrieve high-quality RNA from FFPE tissue efficiently. Various RNA extraction kits for this application are commercially available, but we found the Roche High Pure RNA Paraffin Kit worked the best in our experiments. However, this protocol is tedious and time *consuming*, which may become a limiting step for *high-throughput*.
2. We recommend using several [5, 6] 5-μm sections of FFPE tissue for RNA extraction. The average tissue sections from *paraffin* blocks were approximately 1 cM^2 with the solid tissue in the middle. Our results suggest that 5-pm tissue sections gave better RNA yield than 20-μm sections, probably due to more efficient *deparaffinization* and cell lysis. We also isolated RNA from tissue sections *mounted* on slides.
3. RNA quantitation is an important step to ensure that sufficient material is used in the DASL Assay to *generate* high-quality data. We recommend the *Molecular* Probes RiboGreen Assay Kit for quantitation of RNA samples. The RiboGreen assay can *quantitate* small RNA volumes and measure RNA directly. Other techniques may pick up contamination, such as small *molecules* and proteins. We recommend using a fluorometer, as *fluorometry* provides RNA-specific quantitation, whereas *spectrophotometry* may be affected by DNA contamination, leading to *artificially* inflated amounts.
4. To prequalify RNA samples prior to array analysis, we use a real-time PCR-based method to assess the intactness of the RNA samples. We find this approach more effective than using a combination of RNA quantitation and a gel-based size analysis.

   There are various qPCR methods, many of which could be used for prequalification of degraded RNA samples. We developed the following protocol based on amplification of a fragment of the highly expressed RPL13a ribosomal protein gene (GenBank accession # NM 012423.2), with SYBR Green detection.

   The RPL13a primers (forward primer 5' GTACGCTGTGAAGGCATCAA 3' and reverse primer 5' GTTGGTGTTCATCCGCTTG 3') amplify a 90-base-pair fragment and are designed to span an intron, so they should produce a correctly amplified product only from cDNA but not genomic DNA.

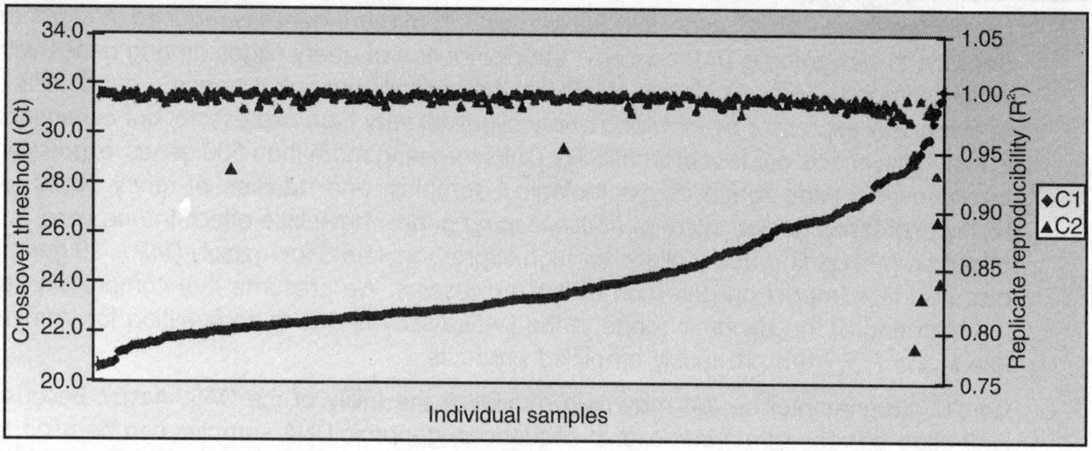

Figure.16.2: Correlation between RNA sample quality control and array data quality. The DASL Assay can effectively generate high-quality gene expression data with samples that have up to eight cycle number differences in qPCR for the highly expressed housekeeping gene, RPL13A (~170-fold difference in "workable" RNA input). Under our qPCR conditions, any RNAs with a Ct number exceeding 28 in the prequalification assay may compromise array data quality.

Using a reaction containing 5 µL of SYBR Green PCR Master Mix, 1 µL of cDNA product diluted 1:10, and 250 nM each forward and reserve primers in a total reaction volume of 10µL, and a PCR profile of 95°C for 12 min, followed by 40 cycles of 95°C for 20 s, 54°C for 2 s, 72°C for 30 s, we routinely measure a crossover threshold (Ct) value of about 17 cycles for intact human reference total RNA (BD Biosciences Clontech, catalog # 636538) using the RPL13a *primers* and the *reagents* listed previously (the detection threshold set at 0.2 for the ABI Prism).

Based on analyses of more than 300 RNA samples derived from formal In-fixed, paraffin-embedded tissues, we found that samples should not exhibit a Ct of more than 28 cycles under these conditions to assure a reasonable expectation of reliable data in the DASL Assay. It is important to recognize that the Ct values we observed are entirely dependent on the equipment and reagents we use.

If other qPCR systems are used, different values may be expected. But, in general, the difference in cycle number at detection between *cryo-preserved* and *paraffin-derived* RNAs should not exceed 8 cycles for good performance in the DASL Assay.

5. DASL Assay query oligos are designed to target intraexonic sequences by default. But the oligos can be designed to span exon junctions for *alternative splicing monitoring*, for example. As a matter of fact, since DASL uses random priming in the cDNA synthesis, the probes can be designed to target any unique regions of the transcripts. There is no need to limit the selection of optimal probes to the 3'-end of the transcripts. The DAP oligo design program can take gene names, RefSeq accession numbers (NM or GI numbers), or sequences. With regard to the number of probes per gene, our analysis has shown that three optimally designed probes performed comparably to four or more probes as to their ability to detect expressed genes as well as *differential* expression in RNA samples. Further lowering the probe number negatively affected assay *reproducibility* slightly, but

people are free to do so, especially if expression from more genes is required. A common question in *designing* a DAP concerns the distribution of query oligos among genes with widely differing expression levels. While it is theoretically possible to swamp out detection of a very low expressor by *including query oligos* for very high expressors, our experience is that these effects are less dramatic. For DAPs targeting more than 500 genes, expression levels cover a wide range across biological samples, and addition of query oligos for highly expressed genes, such as *housekeeping* genes, have little effect. In one case, we observed that adding query oligos for high expressors into a low-plexity DAP (~30 genes) also had little impact on detection of low *expressors*. We presume that compression at the high end of the *dynamic* range in the DASL Assay is due to competition for *priming* late in the PCR from previously amplified products.

6. Careful experimental design may help maximize the utility of the DASL Assay. Because DAP oligo designs target intraexonic sequences, genomic DNA samples can be used to *monitor assay* performance and troubleshoot questionable DASL Assay results for FFPE-derived RNA samples. As an option, one can use the Make SUD (make single-use DNA) process to prepare activated gDNA to include as a positive control sample. The data derived from the gDNA samples can then be used to qualify individual target assays during data analysis in the BeadStudio application.

   A well-designed DASL Assay experiment includes replicates and both gDNA and intact reference RNA positive control samples. We find that the gDNA samples are useful to monitor assay performance in the steps following cDNA synthesis and the reference RNA sample can be used as a positive control relative to the degraded FFPE-derived RNA. *Replicates* can identify those samples that are less robust and less *trustworthy* by lower *correlations*.

7. As mentioned previously, a gDNA sample can be used to qualify the performance of the assay as well as the arrays. Ideally, intensities for all probes should exceed *background* for a gDNA sample. *Exceptions* occur when individual probes are not uniquely mapped to the genome in RefSeq, causing a probe to cross an exon junction. Currently, this occurs for about 1 % of probes, assuming a ~5% rate of genes not mapped in RefSeq coupled with a ~15% rate of randomly targeting an exon junction. In addition, there is the chance of low query *oligo concentration*, leading to low intensities for a few more percent of probes. In our experience, at least 95% of probes should show *intensities* significantly above background for gDNA samples.

8. The data quality for samples with too little input is quite similar to those with too much RNA *degradation*. We observed that poor samples are characterized by a decrease in overall signal across probes, a shift in the distribution of signals to lower *intensities* (that is, fewer genes with high signals), an increase in the intensity of the negative controls, and an increase in the variation among sample replicates at low intensity. To estimate the impact of input RNA quantity on assay performance, we tested various amounts of total FFPE-derived RNA (400 ng, 200 ng, 100 ng, and 50 ng) in the DASL Assay. Each eDNA sample was processed in duplicate for independent *technical replicates*. Highly *reproducible* results were obtained with as little as 50ng of total RNA ($R^2 = 0.97$). However, the larger the RNA input, the more closely the samples correlated, until 200-ng samples

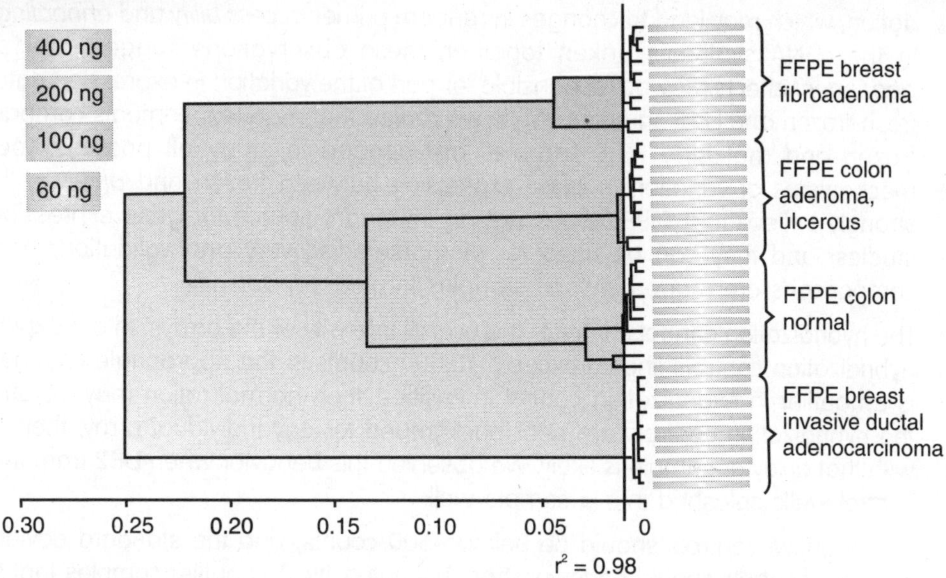

Figure 16.3: The effect of RNA input on assay reproducibility. Lower RNA input increases scatter among technical replicates of FFPE-derived RNAs.

were indistinguishable from 400-ng samples.

9. We previously compared the number of genes detectable by the DASL Assay in RNA samples extracted from paired *fresh-frozen* and FFPE *colon* and *breast* tissues, both cancerous and normal. More than 90% of the genes that were detected in the fresh-frozen samples also were detected in their matching F1-PE samples, when 200ng of total RNA was assayed.

   Lists of differentially expressed genes generated from fresh-frozen and H-PE samples had highly significant overlap (with the FFPE list containing ~50% *fewer genes*), suggesting that sets of differentially expressed genes identified in FFPE samples *resemble* those identified from fresh-frozen samples.

   The correlation between the gene expression profiles of paired fresh-frozen and FFPE tissues was 0.69 on average, but between each pair, this correlation ranged from 0.4 to 0.9. This observation is entirely *consistent* with other measurements of expression results in FFPE vs. *fresh-frozen* tissues, including qPCR data. The result suggests that sequence (or gene)-dependent *effects* may be derived largely from variations in sample *processing* and *handling*.

   It is well known that the quality of RNA extracted from FFPE samples depends largely on multiple parameters: initial tissue fixation, sample processing, storage, and RNA extraction protocol. We believe that the major source of variability in expression profiles using RNA extracted from FFPE tissues is sample quality. As mentioned, we *implemented* a qPCR-based sample prequalification procedure to address this issue. Using this approach, the FFPE samples that showed the lowest *correlations* with matching frozen samples also had the poorest overall *quality*. In addition, RNA *secondary* structures may change after degradation, which may lead to changes in random primer *accessibility* and *annealing* efficiency in the cDNA synthesis. Taken together, these observations suggest that *sequence-dependent* effects may be responsible for part of the variation in expression data between fresh-frozen and FFPE samples. While we realize that more experiments comparing fresh-frozen and matching FFPE samples are needed to study all possible sources and mechanisms of variation in gene expression between frozen and *archived* tissues, we strongly believe that FFPE tissues are an important source for gene expression profiling studies; and they can be used for biomarker discovery and validation, *particularly* if comparisons are made between samples treated *equivalently*.

10. The hybridization controls monitor the overall intensity of the arrays. In our experience, the hybridization controls should exceed -10,000 counts in the appropriate channels. If there is an outlier in hybridization control intensities, then normalization may be attempted. If the hybridization controls are near background for any individual array, then a problem with that array or sample is likely. We observed this behavior when UB2 from the humidity control wells splashed into a sample well.
11. The negative controls should be below ~500 counts, and the standard deviation of the negative controls should be lower than their intensity. Any outlier samples that show high counts or large standard deviations on the negative controls should be considered with caution. We find that high negative controls can be associated with RNAs that are overly degraded and, therefore, unlikely to provide useful expression information. The exception to this rule is in the analysis of gDNA control samples, where we have seen high negative controls occur.

    This phenomenon depends on which genes are targeted in the DAP, presumably due to the increased complexity of gDNA sequences relative to the transcriptome. When high negative controls were observed on gDNA samples, we observed that RNA samples are unaffected.
12. The control summary graph titled "Genes" reports the average intensity and standard deviation of all genes monitored. This value is expected to be lower in RNA samples than gDNA samples and lower in FFPE-derived RNAs than in intact RNAs. Outlier samples with low intensities may reflect overly degraded RNA or very low RNA input. Because the assay monitors the number of amplifiable targets, low RNA input levels can resemble overly degraded RNAs in DASL Assay performance.
13. The gap, stringency, and contamination control query oligos all target the glutaminyl amino tRNA synthase (QARS) transcript. Therefore, the intensities of these controls should reflect the level of expression of the QARS gene in the samples tested. This transcript was chosen because it is a housekeeping gene with medium expression levels in most tissues. The intensities for these controls should show general concordance across samples. The intensities of these controls should follow a general pattern: (1) the gap control should approximate the average intensity of all genes, (2) the intensity in the red channel should dominate the intensity in the green channel for the low stringency control and the opposite should be true for the high stringency control, and (3) only one of the contamination control intensities should exceed background. The way the contamination controls work

is that only one set of query oligos targeting one of four sequences in the QARS cDNA is added to any one tube of DAP. Therefore, only one of the first four columns in the Contamination control graph should show intensity that parallels the *Stringency* and Gap signals across samples. The last four columns indicate intensity on bead types that are targeted only in the GoldenGate Assay for SNP genotyping. If there is a *contamination* with amplicons in the lab, the data should show a large, constant signal across all samples, inconsistent with sample-dependent differences in QARS expression.

14. Another valuable approach to assessing data quality is to cluster the samples using the dendrogram tool in BeadStudio. If there are replicate samples or known sample relationships, outlier samples can be readily detected by their failure to associate with each other in clustering.
15. The DASL Assay combines the advantages of array-based gene expression analysis with those of multiplexed qPCR, thereby offering much higher multiplexing capacity, at high throughput and low cost. It can use as little as 50 ng of total RNA to analyze over 500 genes in FFPE samples, -100-fold less than that required by qPCR, which usually uses 20-50ng per reaction (per gene). Since many genes are measured *simultaneously* in one DASL Assay, it provides an excellent *opportunity* for internal data normalization, thus solving a major problem encountered by qPCR. This becomes very important when cross-sample comparisons are needed, especially when the samples under study have different degrees of RNA *degradation*.

Chapter 17

Open Reading Frames of Genomes

Genome *sequencing* is opening *unprecedented* ways for understanding the primary structure of open *reading* frames (ORFs) on a global scale (ORFeome) and the evolutionary forces that shape them. Codon usage has been intensively studied in many organisms and one already has a relatively good understanding of the structural and functional constraints that shape its evolution. *Conversely*, other important features, such as codon context (two neighbor codons), tandem codon repeats, and amino acid *composition* have not been so well studied and we are still far from understanding their importance for gene stability, mRNA decoding efficiency, and accuracy.

Codon context is rather interesting because it is biased and has an important impact on tRNA decoding accuracy but the rules that define good and bad context of neighbour codons are not yet understood. Additionally, it is not yet clear whether codon context is used to *regulate* speed of mRNA translation, if it influences ribosome drop out during *elongation* and how genes with bad codon context are *translated* under *physiological* stress.

Considering that mRNA decoding accuracy is *critical* to ensure correct flow of genetic information from DNA to protein, understanding those rules is likely to provide new insight on the *constraints* imposed by the mRNA translation machinery on gene evolution. More importantly, codon context rules would allow one to redesign ORFs for optimal gene expression in *heterologous* hosts.

This is of practical relevance because previous studies carried out in our *laboratory* have shown that codon context is species specific and consequently heterologous genes do not have the most *appropriate* context for translation by the host *translational machinery*. Traditional methods for codon usage and context *analysis* do not provide *user-friendly* tools to study gene primary structure on a genomic scale.

Codon usage tables, using *absolute metric*, are available in public databases for any sequenced gene or genome and free-ware software for multivariate analysis of codon and *amino acid* usage is also readily available.

However, sophisticated *statistical* and data *visualization* tools are clearly lacking. To study context bias in complete ORFeomes, we have constructed a *bioinformation* system herein named ANACONDA, which imports FASTA files, and performs a series of *analyses* that permit *elucidating* how codons are associated in consecutive pairs, either in coding sequences or in *noncoding* regions.

This methodology allows to differentiate general biases *imposed* by general rules of genome evolution, which are related to DNA *replication* biases, from biases imposed by the mRNA translational *machinery*. In here, we describe the *architecture* of ANACONDA and how it can be used to *analyze* gene primary structure on an ORFeome scale.

SOFTWARES

ANACONDA is a software package specially developed for the study of genes' primary structure. It reads ORFeomes downloaded from public databases in FASTA format and uses a set of *statistical* and *visualization* methods to reveal information about codon context, codon usage, and *nucleotide* repeats within ORFs. The eneral features of ANACONDA are described below:

1. Software

The ANACONDA software was developed in C++ language with Microsoft Foundation Class, it runs on MS Windows, and can be downloaded for *noncommercial* use from the website: http://*bioinformatics*.ua.pt/aplications/ anaconda.

2. Requirement

Windows (98/Me, NT 4.0 SP6, 2000 or XP), with a 600 MHz Intel Pentium III or equivalent processor and 128 MB of RAM memory (256 MB or more is *recommended*). There should also be 100 MB of available disk space and the minimum resolution should be 800 × 600 (1024 × 768 or higher is recommended).

3. Data

The genomes files processed by the ANACONDA must be in FASTA format.

4. Parameters

The reference values of relative *synonymous* codon usage (RSCU) are necessary for the calculation of the codon adaptation index (CAT). This data must be introduced *manually.* In the current version, the ANACONDA database includes RSCU values for *Candida albicans, Saccharomyces cerevisiae,* and *Escherichia coli.*

EXPERIMENTAL APPROACH

In this section, we describe the main tools of ANACONDA which are divided into four main parts, namely:

(1) uploading and validation of DNA sequence data into the local database,

(2) building ORFeome maps for two-codon context bias,

(3) visualization and analysis of the two-codon context biases in individual sequences, and

(4) comparison of the codon biases across multiple ORFeomes. Also, and taking advantage of the fact that ANACONDA interprets DNA sequences as sequences of *trinucleotides*

(codons), several tools regarding codon usage analysis have been *implemented*, as explained next.

Statistics Methods

ANACONDA uses contingency tables as the basic statistical *methodology* and identifies preferred and rejected codon pairs of an ORFeome through the analysis of adjusted residuals values of the contingency tables. The following list *highlights* the main statistical procedures performed by the software.

1. ANACONDA uploads ORFeome sequences from any genome and reads them in the 5' to 3' direction fixing each codon (ribosomal P-site) and memorizing its neighbour codons (A-site codon).
2. The data extracted by ANACONDA is then transferred to a 64 × 61 contingency table with two *categorical* variables: A and B (2e4). Variable A represents the 64 possible codons located in the ribosome P-site and variable B represents the following (A-site codon) for each observed codon pair in the ORFeome
3. ANACONDA then calculates the value of the Pearson's chi-squared statistic and the adjusted *Pearson* residual values. Pearson's statistic represents a global measure of the difference between observed and expected codon frequencies.

Table 17.1: Contingency Table.

A \ B	AAA	AAC	...	...	...	UUU
AAA	nll	n_{12}	...	n_{1j}	...	$n_{1,64}$
AAC	n_{21}	n_{22}	...	n_{2j}	...	$n_{1,64}$
...						
...	n_{il}	n_{i2}	...	n_{ij}	...	$n_{j.64}$
...						
UUU	n64,1	n64.2	...	n64.,	$n_{64,64}$	

[a] nij is the absolute frequency of the codon pair (Ai,Bj) in the ORFeome, where Ai represents one codon in ribosomal P-site and Bj the following codon in ribosomal A-site.

4. If the hypothesis of independence between the variables A and B, i.e., between contiguous codons, is rejected ANACONDA determines the contributions of each 64 × 61 codon pairs to Pearson's statistic value computing the adjusted residual values.
5. The obtained adjusted residual value, for each pair, is then converted into a gray scale and the information is displayed in a 64 × 61 codon context map, where light gray represents positive adjusted residual values greater than +5 (herein called preferred codon pairs) and the dark gray represents negative adjusted residual values lower than -5 (herein called rejected codon pairs). The adjusted residual values that fall within the interval of -5 to +5 correspond to *codon* contexts that do not contribute to context bias for confidence levels greater than 99% and are shown in black.

PROCESSING DATA

1. Reading sequences

The ANACONDA reads genome sequences stored in FASTA format. The length of each ORF and the number of ORFs in a single file are virtually unlimited. Several files can be opened simultaneously. The imported data, coming from single or multiple files, is classified in a *hierarchical* tree view, considering three different information levels: species, *chromosomes* and *genes*.

2. Validation of ORFs

When scanning the ORFeome, ANACONDA filters pseudogenes or erroneous ORFs resulting from deficient annotation and/or sequencing errors. A number of *quality* controls are defined to allow for filtering the ORFeome. For example, very small ORFs (usually less than 100 nt in length), ORFs whose *nucleotide* sequences are not multiple of three, ORFs without stop codons or ORFs with *premature* stops, are excluded.

Each rule can be individually activated according to user needs. For instance, if the goal is the analysis of all coding and noncoding sequences, all *validation* controls can be deactivated before opening the files.

3. Data processing (quantification)

The imported sequences are then processed according to the statistical *methodology* that reveals the *irregularities* in the codon context along the genome. In this phase, sequence processing can be avoided if the aim is to apply data from a *previous* statistical analysis to a current analysis.

Also, sequences with particular characteristics, or groups of genes can be excluded (at the beginning or at the end) from quantification. The length of the codon context can also be modified, i.e., instead of analyzing *codon-pairs*, triplets of codons or long range context effects can be studied.

4. Evaluation of the sequences quality

Once the raw data is processed, ANACONDA generates a report showing rejected ORFs and a small description of the rejection. Valid ORFs, using particular set of filters, are shown on a specific menu "*Valid Tab*" on the left panel of the screen. ORFs excluded from analysis appear in the "*Rejected Tab*" of the same panel. This allows simple visual inspection of all sequences present in the original FASTA files.

EXPLORING DATA

1. Creating an ORFeome context map

After processing valid sequences, an entry with the species' name, as given by the user, will appear on the left panel of the main window of ANACONDA. This panel follows a hierarchical architecture with individual sequences, *chromosomes*, and *genomes*. Clicking on each group of ORFs, i.e., chromosome or genome, will open the respective map for two-codon context bias on the right panel of ANACONDA's main window

2. Interpretation of genomic maps

The map represents the bias detected for two-codon contexts in the selected set o ORFs. The bias is given by a gray scale in which dark gray stands for rejected and light gray stands for preferred codon pairs, in *relation* to what would be expected in a *non-association* basis.

Each possible *combination* of two codons, i.e., each possible context, is represented by one small square of the map and identified by the codon of the row and the codon of the column to which the small square belongs.

3. Data from individual contexts

To facilitate interpretation and analysis of genomic maps, the two-codon contexts can be selected with the cursor and individual information from them will be displayed in the status bar of the software's window. These include: (1) number of genes used to calculate the bias, (2) full name of both axes of the map, (3) residual value for that context, and (4) occurrence for that codon pair in the genome under analysis.

4. Additional data

Apart from the data directly included in the map, ANACONDA produces additional data about the sequences analyzed, namely:

a. Codon countine and rare codons

The frequency of each codon is *plotted* in a graph, for a chosen set of sequences, either for one chromosome or for the entire genome. This can be obtained with the tool Options ® Rare Codon, because it allows determination of a threshold for codon usage that automatically indicates whether a codon is rare. This window also presents the total number of codons present in all valid ORFs of an ORFeome.

b. Nucleotide counting

Codon context has been further explored focusing on the relative frequency of each *nucleotide* on each position of the neighbour codon, either at the 5' or at the 3' sides. This information is available on dialog Options → Nucleotide Counting that produces a graphical visualization of nucleotide *neighbourhood* for any given codon.

5. Further manipulation of the map

Certain aspects of the map for two-codon context can be altered by the user.

a. Colors and intervalse

The colors used to represent the *deviation* from the expected mean (the residuals scale) can be chosen from a color palette on the Options menu. Also, the residual values defining the different intervals can be modified by the user.

b. Cluster analysis

To define codon context patterns both axes of the 64 × 61 map ca be clustered. Additionally, columns or rows can be ordered alphabetically by the nucleotide at each codon position (N1, N2, or N3). This approach was implemented in response to the *preliminary* observation that some positions from two consecutive codons are highly correlated.

c. Exporting images

The entire map or parts of it can be copied and pasted 'Exporting images into other *applications* (using the drag-and-drop or the edit-copy *functionalities*).

6. Exporting data

The numerical data that give origin to a map can be exported as an Excel worksheet. This will include raw data and residuals data of all map layouts, i.e., 64 × 61 codons, 21 × 21 amino acids, and so on, through the option File → Save Matrix.

HANDLING ORF SEQUENCES

1. Mapping ORFs,

To detect the impact of codon context bias (as well as the presence of rare codons) on coding sequences, ANACONDA has additional tools for sequence mapping. These can be activated by selecting individual ORFs on the *hierarchical* left panel of the software's main window. The layout for sequence analysis (called *"view gene"*) will appear on the main panel and include written information about the ORF and the sequence itself, in which the codons have been colored with the same residual color scale of the ORFeome map.

Again, passing the cursor over the sequences will highlight additional information about each selected context in the status bar of the main window. The threshold for coloring the sequences, together with the choice for *mapping* rare codons on them can be customized by the user at the dialog Options-+View Gene.

2. Exogenous ORFs and codon optimization

To optimize ORF sequences for heterologous gene expression, or for *de novo* gene synthesis, ANACONDA has an algorithm that color codes the sequence of the heterologous ORF according to the codon context rules of the host *expression* system. For this, the user must open the heterologous ORF sequence using the *"no quantification"* option and then redirect the file to the genome of the host of interest. The display window will then show the distribution of good and bad context for that gene.

3. Additional information

Apart from the sequence information shown in the gene view layout, the program offers additional information, obtained from individual sequences or groups of sequences, i.e., chromosomes or total ORFeomes. Selecting the Global gene information option in the View menu the available information about that particular sequence will be displayed. This includes codon and amino acid counting and also several *indexes relevant* for codon usage characterization, such as G + C content at individual codon positions (first, second, or third); the effective number of codons (245; the RSCU value for each codon; and the corresponding CAI.

4. Filters

Searching for specific ORFeome features can be performed using subsets of ORFs. The sequences that comply with the imposed rules are presented in a special tab in the left panel (*Filtered*). The available "filters" include:

(1) searching for special color patterns or codon/amino acid sequences;

(2) searching for runs of up to six rare codons;

(3) looking for ORFs rich in bad contexts or rare codons; and

(4) finding ORFs whose G+C% is included in a chosen interval.

This filter tool is very useful for studying the *distribution* of these variables along an entire ORFeome. It also helps finding specific sequences or ORFs with extreme values for a *particular* variable.

5. Image and data exporting

As with genomic maps, any part of the gene view layout can be selected and copied into another application. Also, *numerical* data associated with filtered ORFs can be exported as Excel worksheet by clicking on the ORF set at the Tab *Filtered* window with the right mouse button.

HANDLING ORFEOME

1. Workspaces

ANACONDA allows the user to work with more than one ORFeome at a time. This creates large data sets that are difficult to deal with, in particular when multiple comparisons are being performed. To overcome this problem, ANACONDA has a Workspaces interface that permits saving all data sets, thus *eliminating* the need of repeating ORFeome *analysis* manually each time one inter-ORFeome analysis is required. When relevant ORFeomes have been opened for the first time the software creates a file of *pathways* that allows ANACONDA to reopen the same at any time.

2. Visualization

All opened files are named as entered by the user, are represented in the hierarchical left panel and sorted by opening order. In this way, each file can be selected, "*navigated,*" and analyzed *independently* as *previously* described

3. Tools for ORFeome comparison

Considering that vast number of ORFeomes can be analyzed simultaneously by ANACONDA, we have included extra tools to allow *comparative* studies.

a. Data normalization

Because adjusted residuals are sensitive to ORFeome size and there is a large size difference between small bacterial and eukaryotic ORFeomes the software includes an option for size normalization that allows direct comparison of all sequenced ORFeomes of the three domains of life.

b. Comparing maps

ORFeome maps for two-codon-context bias can be compared in pairs using the Processing → Compare Genomes option. This tool will produce a differential display map that results from subtracting both maps cell by cell. Differential display maps can also be *manipulated* by the user as described for normal ORFeome maps.

c. Clusterings

Alternatively, all opened maps can be compared in one single display to allow detecting overall patterns of two-codon context. This can be achieved with the option Processing → Compare all genomes. When this option is selected, ANACONDA will transform the 64 × 61 maps of each opened ORFeome into one single column of 3904 lines, one for each *possible* codon pair.

In a second step, all columns are aligned set side by side to allow immediate *comparison* of patterns. As with all 64 × 61 maps, it is possible to rearrange this large-scale comparative map through cluster analysis of both axes to *highlight* major *common* patterns

4. Exporting data

Similarly to the 64 × 61 maps, the adjusted residuals of large-scale comparison maps can be exported as CVS files for further *mathematical* analysis.

POINTS TO REMEMBER

1. The contingency table is a 64 × 61 matrix. Because stop codons do not have codons on their 3'-side the three columns *corresponding* to these three codons are not defined.
2. For a more detailed description of FASTA format *see* www.ncbi.nlm.nih.gov/BLAST/fasta.html. As an example, the complete set of ORFs from a single species can be found in a format appropriate for ANACONDA in .ffn files of GenBank. If needed, this format must be applied to other sequences before opening them with ANACONDA.
3. In most cases, data presented by this software is calculated based on the ORF or ORF set selected in the left panel of the main window. If a special set of ORFs is to be analyzed it must be formatted as a FASTA file containing the chosen ORFs and then be opened by ANACONDA at later stage.
4. In the maps of *two-codon* context created by ANACONDA the rows represent fixed codons as indicated on the left, whereas the columns correspond to codons standing on the 5'- or 3'-sides of the fixed codon, as indicated at the top of the map. The type of context (5'- or 3'-side), as well as the type of map (showing codons, amino acids or *nucleotide* positions), can be chosen using a drop-down menu on the *top-right corner* of the main window.
5. Rare codons are highlighted by a blue circle in the sequence view layout, and will be considered in future versions of ANACONDA as codons to be *preferentially* optimized.
6. Usually, the last position of one codon (N3) is highly correlated with the first position of the following one (N4), as seen by the *formation* of single *colour larger* squares in the maps.
7. By default, when opening a new set of DNA sequences the software will quantify them, i.e., will count codon pairs and calculate the adjusted *residuals*. However, sequences can be opened without quantification to be analyzed with residuals calculated with other sequences. This can be achieved by simply choosing the "No quantification" option of the Processing window.

8. A sequence that has been opened with no quantification can be analyzed with residual data extracted from other Orfeomes. For this, the user must select the sequence using the *hierarchical* left panel and click on its name with the right mouse button. Then the option "*redirect*" must be selected, as well as the genome whose residual data is to be used. The sequence will then appear at the gene view *layout*, colored as if it *belonged* to the host genome.
9. The header of the gene view layout includes: (1) the ORF name, (2) the total number of codons of that ORF, (3) the number of codons whose frequency is below the chosen threshold for rare codons, (4) the *percentage* of rare codons in the ORF, (5) the type of map and how data was quantified to reach the residuals used, and (6) the count and the percentage of two-codon contexts whose calculated residues belong to each color of the scale shown in the layout. Additionally, ANACONDA allows counting the total number of particular codons, as specified in the gene view options.
10. Alternatively, the same information can be obtained using the "i" button of the toolbar. Also, the option View-a Gene (Nc, total GC, GC3, CAI) offers a reduced version of the same information but in a floating window, that allows selecting different ORFs without closing it.
11. The CAI value for the selected sequence will appear only when the RSCU data for reference genes have been typed in. This has to be done manually, choosing Add in the window for defining RSCU values of the Options → Define RSCU Values menu. Each set of RSCU values can be saved for later use. To define the RSCU values of a genome, right button click in the genome name and choose RSCU values: set RSCU values.
12. Some filter tools include an option to *visualize histograms* showing how the variable is distributed across the entire ORF set. For example, to search for a set of ORFs with more than 10% of bad codon context the filter window should be open (either in the "*Processing*" menu or using the button at the toolbar).

    Then the option "*Ratio*" should be selected and the filter for "*Residual Values*" enabled. After choosing the degree of two-codon rejection to search for (according to the residual intervals chosen), and defining the search threshold at 10%, the filter should be run. The same filter can be used in several ORFeomes. However, each time a filter is run a new set of filtered ORFs will be displayed in the "*Filtered*" left panel, eliminating the *previously displayed* ones.
13. *Workspaces* can be named by the user and saved at any *location* in the *file system.*
14. Some windows allow selecting the ORFeome to be analyzed, through a scroll-down menu located in a field called "*genome.*" Usually, the default ORFeome is the first one that was opened, and attention must be taken to change this selection to analyze the intended ORFeome.
15. The adjusted residuals are corrected as if all ORFeomes had the same size, which can be fixed by the user in the Option → *Standardize.*

Chapter 18

Synthesis of cDNA Library

A majority of coding genes that make up an *eanimal's* genome are under selective pressure, such that there is little room for evolving function due to the importance of the wild type gene. However, in some cases, genes that are dormant (in selected regions of *heterochromatin*) or genes that may have arisen as a result of whole or partial gene duplication (with *switching* of selected exons and/or introns), may confer a selective *advantage* to organisms living in a certain environment or exposed to certain conditions.

Extra copies of a gene or "*spliced copies*" may become activated (or mutated) to code for new proteins, in effect supplying "*novel*" genes to the existing suite of genes present in an organism. Although a number of genome projects are complete and *numerous* others are ongoing, a wide variety of *fascinating* organisms exist for which only selected genes have been identified.

By examining the response and regulation of gene expression in systems that not only survive, but, in many cases thrive under adverse *conditions,* insight can be gained into the prevention and potential treatment of these types of *systemic stresses.* The end goal is to determine the function of such genes.

Advances in the field of functional genomics is helping to address this goal and advances in bioinformatics, combined with the *availability* of large data sets for well-annotated genes, has provided a basis for the generation of software and analysis programs that allow *researchers* to examine nucleotide and protein sequences using established "*rules*" and algorithms.

Much information can be gleaned from the nucleotide sequence itself, as well as the primary amino acid sequence of the translated gene, such that informed *hypotheses* can be constructed and tested prior to initiating larger *downstream* projects. Such preliminary analysis sets the stage for further experiments, such as the cloning and expression of open reading *frames,* examination of the upstream promoter, construction of mutated constructs that affect gene expression, and possibly even the creation of a knock-out animal to examine the *in vivo* effects of the loss of select genes. The first step in this process is the *identification* of differentially expressed genes for the condition of interest.

This chapter outlines the basic construction of a cDNA library from a select population of mRNA, including methods for mRNA/cDNA *preparation* and *cloning,* subsequent screening for differen-tially expressed genes, and *downstream confirmation* via Northern blotting.

Methods for analysis of select sequences are also outlined using the previously unreported novel gene, *sarp-2,* which is included as an example. *Sarp-2* is one of several novel clones isolated from a cDNA library constructed with mRNA prepared from *hepatopancreas* of the marine snail, *Littorina littorea;* the gene is induced by anoxic exposure. Identification and characterization of this novel gene was performed as outlined in this chapter and the results are reported in the *accompanying* figures.

REQUIREMENTS

All chemicals used are of molecular biology grade or their equivalent and of the highest *purity.* All plastic and glassware, including bottles and pipet tips, are *autoclaved* and gloves must be *worn* at all times during operations involving *nucleic acid* manipulation.

Obtaining Pure RNA

1. RNase-free water. Add 1 mL of diethyl pyrocarbonate (DEPC) (Sigma-Aldrich) to 1 L of water (0.1% v/v), stir overnight (>12 h), autoclave. This destroys any RNases present and this water will be used to make up solutions in this section and to dissolve RNA samples.
2. *Trizol reagent* (Invitrogen).
3. *Chloroform* (Fisher Scientific).
4. *Isopropyl alcohol* (Fisher Scientific).
5. 70% *Ethanol.* Add 30 mL of DEPC-treated water to 70 mL of *ethyl alcohol* (200 proof; Phannco).
6. Oligo(dT) cellulose (New England BioLabs). Dry oligo(dT) cellulose is combined with 0.1 *M* NaOH to create a slurry and poured into a sterile column or 1-mL *syringe plugged* with sterile cotton or glass *wool.* The column is *equilibrated* with loading buffer prior to adding sample.
7. Loading buffer: 1 *M* NaCl, 2 mM *phosphate* buffer, pH 7.2.
8. Middle wash buffer: loading buffer + 0.3 *M* NaCl.
9. *Elution buffer*: 10 mM Tris-HCl pH 7.2-7.4, 1 mM EDTA.
10. 3 *M* sodium acetate, pH 5.2.
11. *Ethanol* (200 proof; Pharmco).

Synthesis of cDNA Library

1. Cloning vector for cDNA library synthesis. Many vec..ors exist, often designed for a specific application. Researchers should examine the 'eatures of those available and decide on a vector appropriate for their application. Some common vectors used in cDNA library synthesis include: Uni-ZAP XR (Stratagene), pITriplEx2 (Clontech), pSPORTI (Invitrogen), 1SCREEN-1 (Novagen/Merck). The Uni-ZAP XR cloning vector will be used for illustrative

purposes in this chapter.

2. 1 μg of column purified mRNA.
3. Oligonucleotide linker-primer containing a poly (dT) region. (ex. 5'-NNNNN NNNCTCGAGdT(15)-3').
4. RNasin (20 U/pL; Promega).
5. AMV reverse transcriptase (10 U/μL) with 10 × reverse transcriptase buffer (Prornega).
6. Nucleotides (dATP, dGTP, dTTP; 10 m*M* each dNTP)
7. 5-Methyl cytosine analog (*dCTP; 5 m*M*).
8. Ribonuclease (RNase) H (5 U/pL; New England BioLabs).
9. DNA polymerase 1 (*E. coli,* 10 U/μL) and 10 × DNA polymerase 1 buffer (New England BioLabs).
10. *Eco* RI adapters. An *EcoRI* adaptor can be purchased commercially or constructed using two primers of different lengths, which can be *hybridized* to form a duplex with a 5' phosphate on the blunt end. Include an overhang of 5'-AATTCNNNN-3' on one primer to create an *EcoRI* restriction site. The opposite end of the duplex (the blunt end of the adaptor) can be ligated to the ends of any cDNA containing a 5' phosphate group. The adapters should be diluted in sterile ddH_2O to a final concentration of 5 μg/μL.
11. 50 × TAE buffer: 2 M Tris-acetate (Tris + acetic acid) pH 8.5, 100 m*M* EDTA.
12. 1 % TAE agarose gel: 1 × TAE buffer, 1 % agarose (1 g/100 mL), ethidium bromide (1 pg/mL).
13. 100 mM EDTA.
14. T_4 DNA ligase (400 U/μL) and 10 × T_4 DNA ligase buffer (New England BioLabs).
15. T_4 polynucleotide kinase (10 U/μL) and 10 × T_4 polynucleotide kinase buffer (New England BioLabs).
16. TSM buffer: 50 mM Tris-HCl, pH 7.5, 100 m*M* NaCl, 8 $_{m}M$ $MgSO_4$, 0.01 % (w/v) gelatin.
17. XL1-Blue (Stratagene).
18. LB-ampicillin broth: 10 g of tryptone, 5 g of yeast extract, 5 g of NaCl, pH 7.5. Dilute up to 1 μL with ddH_2O and autoclave. Broth is supplemented with 0.2% (w/v) maltose and ampicillin (100 pg/mL final concentration) after broth has cooled to room temperature.
19. NZY agar: 5 g of NaCl, 2 g of $MgSO_4 7H,O$, 5 g of yeast extract, 10 g of peptone, 15 g of agar, pH 7.5. Dilute to 1 L with ddH_2O and autoclave. Pour plates when agar has cooled to approx. 50°C and store at 4°C until used. 'the agar can also be stored at 4°C and remelted in a microwave just before use.
20. NZY top agar: 5 g of NaCl, 2 g of $MgSO_4 7H_2O$, 5 g of yeast extract, 1 g of peptone/L, 0.7% w/v agarose, pH 7.5. Dilute up to 1 L with ddH_2O and *autoclave*. The top agar is stored at 4°C and remelted in a *microwave* just before use. The agar must be cooled to <45°C before addition of *bacteria*.
21. *Pfu* polymerase (2.5 U/μL) (Stratagene).

Amplification of the Library

1. NZY plates (24 × 24 cm), LB-*ampicillin broth*, XL1-Blue cells, NZY agar, NZY top agar.
2. TSM buffer.
3. Chloroform (Fisher Scientific).
4. Dimethyl sulfoxide (DMSO; Sigma-Aldrich).

Screening of the Library

1. NZY plates (24 × 24 cm; see Subheading elsewhere in this chapter).
2. Hybond-N 0.45-μm nylon membranes (GE Healthcare).
3. 20× SSC: 3 M NaCl, 0.3 *M* Na-citrate (for 1 L: 175 g of NaCl, 88 g of sodium citrate); dilute as necessary with ddH_2O.
4. Denaturation buffer: 1.5 *M* NaCl, 0.5 *M* NaOH.
5. Neutralization buffer: 1.5 M *NaCl,* 0.5 M Tris-HCl, pH 8.0.
6. Rinse buffer: 0.2 M Tris-HCl, pH 8.0, 2× SSC.
7. Whatman filter paper or chromatography paper (Fisher Scientific).
8. dNTPs: dATP/dTTP/dGTP (10 mM each dNTP).
9. Anchored Oligo(dT) primer: (5'-dT(15)(A/G/C)-3'). Can be purchased commer-cially (Invitrogen; New England BioLabs).
10. RNasin (20 U/μL; Promega).
11. Dithiothreitol (DTT) (Sigma-Aldrich). Make up a 0.1 M stock with sterile ddH_2O.
12. AMV reverse transcriptase (10 U/μL) with lOx reverse transcriptase buffer (Promega).
13. [α-^{32}P]dCTP (3000 Ci/mol; GE Healthcare).
14. RNase A (60 mg/mL in ddH_2O; Sigma-Aldrich).
15. Quick Spin columns (TE; Sephadex G-25, Fine) for radiolabeled DNA purification (Roche).
16. 1 × T_{10} E 1 buffer: 10 mM Tris-HCl, pH 8.0, 1 mM EDTA.
17. Hybridization buffer. Modified Church's Buffer: 0.25 M Na_2HPO_4, 0.25 M NaH_2PO_4 (pH 7.5), 1 mM EDTA, 7% sodium dodecyl sulfate (SDS; w/v); or buffer can be purchased commercially: Ultrahybe (Ambion).
18. Membrane washing buffer: 0.1 × SSC, 0.1 % SDS (w/v).
19. Membrane rinsing buffer: 0.l × SSC.
20. X-ray film (Kodak).
21. TSM buffer.
22. Chloroform (Fisher Scientific).
23. XL1-Blue culture.
24. NZY top agar.

In Vivo Excision of Clones

1. XLI-Blue cells.
2. 10 mM $MgSO_4$.
3. ExAssist helper phage.
4. Uni-ZAP XR phage stock.
5. LB-ampicillin broth.
6. SOLR cells.
7. LB-ampicillin agar plates: Agar (15 g) is added to LB broth (1 L) and autoclaved. Cool to <50°C, add *ampicillin* (100 µg/mL), and pour into 10-cm plates at desired thickness.

Miniprep Plasmid Isolation

1. LB-ampicillin broth.
2. Prelysis buffer: 50 mM glucose, 25 mM Tris-HCl pH 8.0, 10 mM EDTA.
3. RNase *A*.
4. Alkaline lysis buffer: 0.2 N NaOH, 1 % SDS. This buffer must be made fresh just prior to use.
5. Neutralization buffer: *5 M* potassium acetate, 30% acetic acid, 50 MM glucose, 25 mM Tris-HCl, pH 8, 10 mM EDTA.
6. Isopropanol (Fisher Scientific).
7. 70% Ethanol.

Role of Restriction Enzymes

1. Restriction enzymes: *Eco*RI and *Xho*I (New England Biolabs); these may be adjusted *depending* on the *adaptors* ligated.
2. 10× restriction enzyme buffer. Buffer will depend on restriction enzymes being used and is supplied with the enzyme.
3. DNA loading dye: 0.25% w/v bromophenol blue, 0.25% w/v *xylene cyanol FF*, 50% v/v glycerol.
4. 1 % TAE agarose gel and 50× TAE buffer.
5. DNA ladder (*Invitrogen*). A ladder should be chosen based on the size of the clones expected (*ranging* from 100 by up to several kilobases).
6. Filtered pipette tips (1 mL).
7. Sterile cotton or glass wool.

Synthesis of Probe

1. dNTPs without dCTP (dATP/dTTP/dGTP; 5 mM each).
2. Random primer (100 mM $d(N)_6$; New England BioLabs).
3. 10× random primer buffer: 0.5 *M* Tris-HCl, pH 7.6, 20 *mM* DTT, 50 *mM* $MgCl_2$, 0.4 M KCl.

4. Klenow fragment of DNA polymerase 1 (5 U/μL; New England BioLabs).
5. [α-^{32}P]dCTP (3000 Ci/mol; GE Healthcare).
6. Quick Spin columns (TE; Sephadex G-25, Fine) for radiolabeled DNA purification (Roche).

Method of Northern Blotting

1. 10× MOPS buffer: 200 mM 3-(N-Morpholino)propanesulfonic acid (MOPS), 50 *mM* sodium acetate, 10 mM EDTA, pH 7.
2. 1.25% Agarose formaldehyde denaturing gel. Melt 3.75 g of agarose in 217 mL of ddH_2O containing ethidium *bromide* (1 μg/mL) in a sterile flask. Place solution in an *incubator* set to 60°C. Into a separate sterile flask, add 30 mL of MOPS 10× buffer and 53 mL of *formaldehyde* 37% (v/v) and place this solution at 60°C. Once both solutions have equilibrated to 60°C, combine the contents of both *flasks* together, gently *swirl* without introducing bubbles, and pour into large gel tray to desired thickness.
3. RNA sample buffer: lx MOPS buffer, 2.2 *M formaldehyde,* 50% (v/v) *formamide.*
4. RNA loading buffer 6× stock: 1 × MOPS buffer, 50% (v/v) formamide, 40% (v/v) glycerol. Add a few flakes of *bromophenol* blue and *xylene cyanol* as *tracking* dyes.
5. Hybond 0.45-pm *nylon* membrane (GE Healthcare).
6. 20× SSC. Dilute as necessary with ddH_2O.
7. Whatman filter paper or *chromatography* paper (Fisher Scientific).
8. Paper towels.
9. Flat piece of *Plexiglas* or sequencing gel plate.
10. Solid weight.
11. Glass or plastic pipet (5- or 10-mL) for smoothing the gel/membrane/filter paper and removing air bubbles from transfer set-up.
12. Plastic wrap.
13. Northern blot fixing solution: 0.05 *N* NaOH.

Northern Blots

1. Hybridization solution: Ultrahybe Hybridization buffer (Ambion) or modified Church's buffer.
2. Radiolabeled probe.
3. Northern blot washing solution: 0.l × SSC, 0.1% SDS in sterile ddH_2O.
4. X-ray film or *phosphoirnager* screen.
5. Northern blot *stripping* solution: boil a solution of ddH_2O and 0.5% SDS and immerse blots to be stripped. Repeat as necessary until *radioactive* label (cpm) detected using a handheld Geiger counter is at or below background levels.

METHODS

Commercial products have been designed for specific types of experiments (high vs. low copy number genes, etc.) that *examine* and identify differen-tially expressed genes. Although the *technology* changes rapidly, cDNA library construction using mRNA isolated from a *tissue/organ/organism* of interest remains the standard method for identifying new and *novel genes* and proteins.

Synthesis of such a library can be carried out using a wide variety of available products; a common standard method for library construction, *screening* and positive clone analysis is outlined in this chapter. These *instructions* assume the use of the Uni-Zap-cDNA synthesis kit. but can be adjusted for the use of other vectors.

Some common phagemid cloning vectors available include, but are not limited to: pITriplEx2 (Clontech), pSPORT 1 (Invitrogen), and 1SCREEN-1 (Novagen). Following *construction* of the library, differential screening will identify selected clones that are induced (or repressed) in the *condition* being tested. In the example provided here, we follow the steps used to find and characterize the novel gene *sarp-2,* which is induced in the *hepatopancreas* of *Littorina littorea* in response to anoxia exposure.

This up-regulation is confirmed with Northern blotting. Following sequencing of the *sarp-2* clone, sequence analysis can be performed using the applications outlined in described any where in this chapter.

Extraction of RNA

1. A number of kits are available today for total and mRNA isolation (Ambion, Roche, Qiagen, Invitrogen, Novagen, etc.). These kits often provide high quality, pure RNA, but are relatively expensive. A *traditional* method for RNA *isolation* is outlined in this section.
2. Homogenize tissue samples in *Trizol reagent* (1 mL per 100 mg of tissue) and add chloroform (0.2 mL/mL of reagent). Invert samples repeatedly (15 s), incubate at room *temperature* (2-3 min), and *centrifuge* at 12,000g at 4°C for 15 min.
3. Remove the aqueous layer and transfer to a sterile Eppendorf tube. Add an equal volume of *isopropanol* to precipitate the RNA. *Incubate* samples at room temper-ature for 10 min.
4. Pellet the RNA by centrifuging the samples at maximum speed in a microcentrifuge for 10 min at 4°C. Aspirate the supernatant and wash the pellet with 70% ethanol (250 μL). Recentrifuge the RNA at maximum speed for 5 min. *Aspirate* the *ethanol* and air-dry the pellet for 10 min on the lab bench. Dissolve the RNA in an appropriate amount of DEPC-treated water (*assuming* you ge(10 pg for every 10 mg of tissue) and store at 4°C (if it will be processed within a week) or at -20°C (long-term storage).
5. Determine RNA *concentration* and *purity spectrophotometrically* by measuring *absorbance* at 260 and 280 nm. For RNA of acceptable quality, the A_{260}/A_{280} ratio should fall between 1.6 and 2.0.
6. Purify the poly(A)+ mRNA from total RNA samples using oligo(dT) cellulose. *chromatography.* Heat total RNA to 65°C, mix with *loading* buffer (1:1 v/v), and load onto an *oligo*(dT) *cellulose* column.
7. Wash the column with loading buffer (1 bed volume) and collect the eluate in a sterile *Eppendorf tube.* Load the recovered eluate onto the column, collect it as it *elutes,* then reload onto the column (repeat twice).

8. Wash the column with loading buffer (1 bed volume), followed by middle wash buffer (1 bed volume), and finally with elution buffer (1.5 bed volume). This last step elutes the poly(A)$^+$ mRNA, to which 0.1 volumes of 3 M sodium acetate, pH 5.2, and 2.5 volumes of absolute ethanol are added. The poly(A)$^+$ mRNA is placed at –20°C and left to precipitate overnight. The *protocol* can be *stopped* here.
9. The following day, centrifuge the tubes at 13,000g for 45 min at 4°C. Discard the *supernatant* and wash the pellet with 70% ethanol (250 ML) and *centrifuge* at the same speed for 10 min. The *resulting* pellet consists of mRNA, which is air-dried and resuspended in DEPC-ddH$_2$O. If the mRNA is not going to be used *immediately,* store at -80°C until further use.

Synthesis of cDNA Library

1. Combine mRNA (5 µg; <30 µL) with the *oligonucleotide* linker-primer (3 µg) and RNasin (5 µL) in an RNase-free Eppendorf tube. Synthesize a complementary first strand of eDNA, using the mRNA as template, with the addition of dNTPs (3 µL), 5-methyl-dCTP (3 µL), lOx reverse *transcriptase* buffer (5 µL), and adjusting the total volume up to 45 ML. Mix the contents gently and allow primer and template to anneal at room temperature for 10 min.
2. Add reverse transcriptase to the reaction tube (5 µL; 50 U) to bring the final volume up to 50 µL. Gently mix the contents of the tube, briefly *centrifuge* to collect contents at the bottom of the tube, and then place at 42°C for 1 h.
3. Second-strand cDNA synthesis (complementary to the first strand) is carried out in the same tube. The following components are added to the first strand reaction: dNTPs (6 µL), 10× DNA polymerase 1 buffer (20 µL), [α-^{32}P]dCTP (1 µL), RNase H (1 µL: 5 U), DNA polymerase 1 (10 µL: 100 U). Adjust the final reaction volume to 200 µL with sterile ddH$_2$O and incubate for 2 h at 16°C.
4. Place the newly double-stranded cDNA reaction on ice and blunt-end the *termini* by adding 22.5 µL of dNTPs (A/G/C/T) and *Pfu* DNA polymerase (2.5 ML) into the tube and incubate at 72°C for 30 min.
5. Add an equal volume of phenol-chloroform (1:1 v:v, pH 7.4) and vortex-mix the tube briefly. Centrifuge the sample at maximum speed in a desktop centrifuge for 10 min at room temperature. Transfer the aqueous layer to a fresh Eppendorf tube and add an equal volume of chlorofonn. Vortex-mix the tube briefly and centrifuge as above for 5 min. Transfer the aqueous layer to a fresh tube.
6. Add an equal volume of isopropanol to the sample, followed by addition of sodium acetate (to a final concentration of 0.1 M). Place the tube at -20°C overnight to allow the cDNA to precipitate. The protocol can be stopped at this step.
7. The following day, centrifuge the *precipitated* cDNA for 1 h at maximum speed in a microcentrifuge at 4°C. Transfer the supernatant into a new tube and store until cDNA precipitation has been confirmed. Using a *Geiger* counter, confirm radiolabel incorporation into the precipitated cDNA pellet. Check the supernatant to confirm that most of the *radioactivity* is in the pellet.
8. Wash the pellet by adding 70% ethanol (250 µL), followed by gentle agitation. Centrifuge

the tube at maximum speed in a microcentrifuge for 10 min. Carefully remove and discard the ethanol, taking care not to disturb the pellet, and air dry for 10 min at room temperature. The location of the pellet should be marked on the outside of the tube with a labelling marker because it will "disappear" once it is dried.

9. Resuspend the cDNA in sterile ddH_2O (10 µL). Gently flick the sample and briefly centrifuge to collect contents at the bottom of the tube. The tube is kept on the bench for 10 min at room temperature (or at 4°C for 30 min) to allow the eDNA to fully dissolve. Transfer the cDNA to a new tube and check the old tube with a Geiger counter to confirm cDNA was completely resuspended.
10. Ligate *EcoRI* adapters (1.5 µL) to the blunt-ended termini by adding T_4 DNA ligase (5 U) and 10× T_4 DNA lipase buffer (1.5 µL). Final reaction volume is 15 pL (make up the final volume with sterile ddH_1O if required). Incubate the reaction for 16 h at 10°C.
11. Heat inactivate the ligase by placing the tube at 75°C for 1.5 min, followed by brief centrifugation to collect the contents at the bottom of the tube. Place the sample on ice and add the following components: T_4 polynucleotide kinase (10 U), 10×T_4 polynucleotide kinase buffer (0.5 µL). Adjust the final reaction volume in the tube to 20 µL. Incubate the reaction at 37°C for 30 min, followed by heat inactivation of the kinase at 75°C for 15 min. Centrifuge the tube briefly to collect the contents at the bottom and place tube on ice.
12. Digest the linker primer's *XhoI* site by adding 4 µL of the appropriate 10× restriction enzyme buffer, 10 µL of sterile ddH_2O, and 6 µL of *XhoI* enzyme to the cDNA (to a final volume of 40 µL). Mix the reaction by flicking, centrifuge briefly to collect contents at the bottom of the tube, and place at 37°C for 2 h. Add an equal volume of isopropanol and place tube at -20°C overnight to precipitate the digest.
13. Centrifuge the precipitated cDNA at maximum speed in a microcentrifuge for 1 h at 4°C. Aspirate and discard the supernatant, taking care not to disturb the pellet.
14. Wash the pellet with 70% ethanol (250 µL) and centrifuge for 10 min at maximum speed at 4°C. Carefully aspirate the supernatant and airdry the pellet on the benchtop for 10 min. Resuspend the pellet in the appropriate volume of ddH_2O (in this case we will use 50 ML) and store at 4°C.
15. Prepare 10 mL of 1% TAE agarose gel, pour into a 10-cm plate and allow the gel to polymerize. Create a grid on the bottom of the plate with 10 squares.
16. Using a 100 mM EDTA solution, make serial dilutions (from 10 ng/µL up to 250 ng/µL) of a DNA standard with a known concentration (such as the DNA ladder described eslewhere in this chapter).
17. Label a square for each of the serial dilutions (0. 5. 10. 25. 50. 75. 100. 150, 200, 250 ng/µL).
18. Carefully pipet 1 µL of each dilution onto the surface of the plate.
19. Below each dilution, spot 1 µL of the cDNA sample and allow all the liquid to absorb into the plate for approx. 20 min.
20. View the plate using ultraviolet (UV) light and estimate the concentration of the cDNA

sample by comparing with the standards.

21. Combine an aliquot of the resuspended cDNA (~100 ng) with the vector arms (1 μg), 1 OX DNA ligase buffer (1 μL), T_4 DNA ligase (2.5 U), and sterile ddH_2O up to 10 μL. Incubate the reaction for 16 h (overnight) at 10°C.
22. The ligation is packaged into phage using a commercially available packaging kit, compatible with the phagemid vector being used, according to the manufacturer's instructions (in this example, GigapackIII XL from Stratagene would be used). Experimental DNA is combined with the packaging extract, mixed and incubated for a set amount of time, chloroform separated, and then the phage is ready to be titered.

Amplification of the cDNA Library

1. From a stock vial, streak XL1-Blue cells onto an LB-ampicillin agar plate and incubate overnight at 37°C. The following day, prepare 25 mL of LB-ampicillin broth culture by inoculating a single colony into a flask and incubating on a shaker at 37°C until the OD_{600} = 1.
2. Centrifuge the XL1-Blue culture at 1000g for 10 min at 4°C. Resuspend the cell pellet in sterile 10 $_mM$ $MgSO_4$ and adjust the volume until the OD_{600} is 0.5.
3. Prepare aliquots (100 μL each) of serial library dilutions (1, 1:10, 1:100, 1:1000, 1:10,000) in TSM buffer, mix each with an equal volume of freshly cultured XL1-Blue cells, and incubate at 37°C for 20 min.
4. Mix the contents of each tube with NZY top agar (~3 mL for each dilutison; cooled to <50°C) and immediately spread on l0-cm plates. Allow plates to cool (until top agar has solidified), then invert and incubate at 37°C until plaques are observed to form. Count the number of plaques on each plate and determine library titer using the following formula: pfu/mL = (number of plaques × dilution factor)/volume of extract plated.
5. To amplify the cDNA library, prepare 10 NZY plates (24 × 24 cm) and an overnight LB-ampicillin broth culture of XL1-Blue cells (as described in steps 1 and 2 of this subheading).
6. Pipet XL-1 Blue cells (100 μL) into 10 sterile 1.5-mL tubes, add an aliquot of packaged phage mixture (containing ~50,000 phage particles) to each tube, and incubate for 15 min at 37°C.
7. Combine the contents of each tube with 7 mL of molten NZY top agar and spread evenly onto each of the NZY plates.
8. Incubate the plates at 37°C overnight to allow plaque formation on the bacterial lawn. Following plaque formation, spread TSM buffer (10 mL) over the surface of each plate, followed by incubation at 4°C overnight with gentle shaking to release the phage from the solid media and into solution.
9. Pool the TSM media containing bacteriophage from the 10 plates and add chloroform to 5% v/v. Incubate this mixture at room temperature for 15 min, centrifuge at 500g for 10 min, then transfer the supernatant to a new sterile container.
10. Add DMSO to the supernatant (7% v/v final) and aliquot the amplified library into sterile

1.5-mL tubes and store at -80°C. The titer of each amplified library should be determined as described above.

Screening of the cDNA Library

1. Prepare NZY plates (10 plates; 24 × 24 cm) as described elsewhere in this chapter. Adjust the amount of phage library used to infect the bacteria using the calculated titer, so each plate contains approx. 20,000-50,000 pfu.
2. Plaques will develop over the course of approx. 12 h and should be viewed regularly until they are round in shape and measure approx. 1 mm across.
3. Place a Hybond-N nylon membrane onto each plate containing the newly developed plaques for one min (adsorption time). Orientation marks should be made on the membrane and on the plate so that the correct plaques can be chosen when going back to select positive clones.
4. Soak each membrane in denaturation buffer for 1 min, neutralization buffer for 5 min, and rinse buffer for 0.5 min.
5. Allow membranes to air dry on filter paper and UV cross-link the phage DNA to membranes using an ultraviolet (UV) crosslinker.
6. Repeat steps 3-5 using a second membrane on each plate, with the adsorption time lengthened to 2 min.
7. Bake the plaque lifts in a gel dryer for l h at 80°C between two fresh pieces of filter paper.
8. Screening requires the synthesis of two sets of probe. One set is synthesized using control mRNA as a template and a second set using mRNA from the stress or condition being examined.
9. Synthesize probe using a protocol similar to that used for first strand cDNA synthesis. Aliquot poly(A)$^+$ mRNA template (about 1 gg) into an autoclaved, DEPC-treated, 1.5-mL microfuge tube and heat at 65°C for 5 min.
10. Cold-snap the mRNA on ice, with subsequent addition of lOx reverse transcriptase buffer (5 µL), dNTPs (3 ML, without dCTP), oligo(dT) primer (3 µg), RNasin (5 µL) into each tube. Adjust the total volume up to 40 µL. Flick the tubes gently to mix, centrifuge briefly to collect contents at the bottom of the tube, and allow the primers to anneal to the RNA at room temperature for 10 min.
11. Add 5 µL of reverse transcriptase and 5 µL of [a*-^{12}P]dCTP (3000 Ci/mol) to each tube and incubate at 42°C for 1 h. Transfer the tubes to a water bath preset to 16°C.
12. To degrade the RNA, add RNase A (5 µL) and incubate at 37°C for 30 min. This will leave a radiolabeled first strand of cDNA, which is loaded onto a quick spin column and centrifuged at 400g for 3 min. The eluant (labeled cDNA) is used to probe membranes.
13. Plaque lift membranes are pre-hybridized in hybridization solution (enough to cover the membrane) for 30 min in a rotating hybridization tube at 55°C.
14. Boil the radiolabeled probe for 5 min and place immediately on ice. Add probe directly to the existing hybridization solution to a concentration of l × 10^6 cpm/mL. To allow for

direction comparison, add identical levels of radioactivity (cpm) to paired blots (control vs. experimental). Hybridize blots overnight (~16 h) with rotation at 45°C.

15. After hybridization, wash the plaquelift membranes in 0.2× SSC, 0.1 % w/v SDS at 55°C for 10 min. Repeat this wash step 3 additional times to reduce background radioactivity. If background levels remain high (as detected using a handheld Geiger counter), washes can be repeated as desired.
16. Place washed membranes radioactive side down on plastic wrap while still damp and seal tightly so that liquid does not leak out. Wrap the membranes a second time using a plastic wrap and smooth the surface using a paper towel (to remove wrinkles or air bubbles). Expose the blots to X-ray film (or a phosphoimager screen) for an appropriate length of time. Exposure time depends on the cpm of the specific probe bound to each membrane and should he adjusted accordingly.
17. Develop the X-ray film (or phosphoimager screen) and analyze the resulting pattern of spots on the autoradiograms. An example of a primary screen autoradiogram is shown in Figure elsewhere in this chapter. Each spot represents a plaque where probe has hybridized to cDNA.
18. Using the orientation markers, trace differentially expressed clones back to their corresponding phage plaques on the NZY plates and excise. A sterile Pasteur pipet is used to excise each section of agar, which is then placed into a 1.5-mL Eppendorf tube containing 500 µL of TSM buffer. If it is not possible to isolate a single clone (especially common during the primary screen) putatively positive "regions" can be excised.
19. Add chloroform to each tube (to 5% w/v), vortex-mix for 30 s, and leave at room temperature for 30 min, allowing the phage to desorb from the agar into the buffer. 20. Centrifuge samples for 5 min at 3000g at 4°C and transfer the supernatants (containing isolated phage) to fresh Eppendorf tubes and store at 4°C.
21. Perform a spot titer for each putative clone or "region" to determine the approx. pfu/mL.
22. Re-plate each putative positive clone or "region" on a separate circular (10 cm diameter) NZY plate at 50-100 pfu/plate.
23. Construct plaque lifts for this secondary screen using small, circular nylon membranes and follow the same steps outlined above for a primary screen (starting at step 3). Repeat differential screening in an identical manner (starting at step 8.). Select positive clones representing genes of interest (differentially expressed) by carefully examining the resulting autoradiograms. An example of a secondary screen autoradiogram is shown in Figure elsewhere in this chapter.

In vivo Excision of Clones

1. The procedure followed in this section depends upon the cDNA library vector used. Components must be purchased commercially (generally in the form of a kit from the same manufacturer that created the vector). A brief overview of the procedure used with the Uni-Zap cDNA libraries (ExAssist Helper Phage with SOLR cells; Stratagene, for details see elsewhere in this chapter) is provided.

2. Prepare a stock of XLI-Blue cells as outlined in elsewhere in this chapter Prepare a stock of SOLR cells using the same protocol as that for XLI-Blue cells. Cells are grown at 37°C in LB-ampicillin broth overnight (~16 h) and are centrifuged at 3000g for 5 min.
3. Discard the supernatant and resuspend each bacterial cell pellet in 10 mM $MgSO_4$ to an OD_{600} = 1.0. Store the SOLR cell pellet at 4°C until step 8.
4. Aliquot 200 pL of each phage stock into a separate Eppendorf tube, add an equal volume of XLI-Blue Cells, and 1 μL of helper phage.
5. Incubate the tubes at 37°C for 15 min. Prepare an equal number of 15-mL tubes containing 3-5 mL of LB-ampicillin broth. Following the incubation, transfer the contents of each Eppendorf tube into labeled 15-mL tubes. Incubate these tubes at 37°C with gentle shaking for approx. 3 h.
6. Centrifuge the tubes for 15 min at 400g to pellet cells and debris and transfer the supernatant, containing the intermediate filamentous phage, to a fresh tube and incubate at 70°C for 15 min.
7. Centrifuge the new tubes for 15 min at 3000g to pellet any remaining cells, and then transfer the supernatant containing the filamentous phage stock to fresh 15-mL tubes and store at 4°C.
8. Add filamentous phage stock (100 μL) representing each selected phagemid into labeled tubes. Resuspend SOLR cells with brief mixing and aliquot 200 μL into each tube and incubate at 37°C for 15 min.
9. Pipette an aliquot of cells (100-200 μL) from each tube onto an LB-ampicillin plate and incubate at 37°C overnight.
10. Streak a single colony from each plate onto a fresh LB-ampicillin plate to create a master plate of SOLR cells containing the pBluescript plasmid of interest. Incubate these plates overnight at 37°C.
11. Inoculate a single colony from a master plate into 3 mL of LB-ampicillin broth. When the OD_{600} = 1.0, a glycerol stock is created by combining an aliquot of the log-phase bacterial culture with sterile glycerol (15% v/v final concentration). Each tube is mixed and kept at -4°C for 30 min, then stored at -80°C.

Separation of Miniprep Plasmid

1. A variety of kits are commercially available that provide a quick and clean plasmid isolation. Such kits are available from a number of companies including, but not limited to Nucleospin (BD Biosciences), Qiagen, and Invitrogen. A standard "non-kit" protocol is outlined here.
2. Select a single colony from a master plate and inoculate 3 mL of LB-ampicillin broth in a 15-mL tube. Incubate overnight (~16 h) at 37°C with shaking.
3. Pipet 1.5 mL of the overnight culture into an Eppendorf tube and centrifuge at 3000g at 4°C for 10 min. Aspirate and discard the supernatant.
4. Resuspend the pellet in 400 μL of prelysis buffer, transfer to a sterile Eppendorftube, add RNase A (5 μL) and vortex-mix.

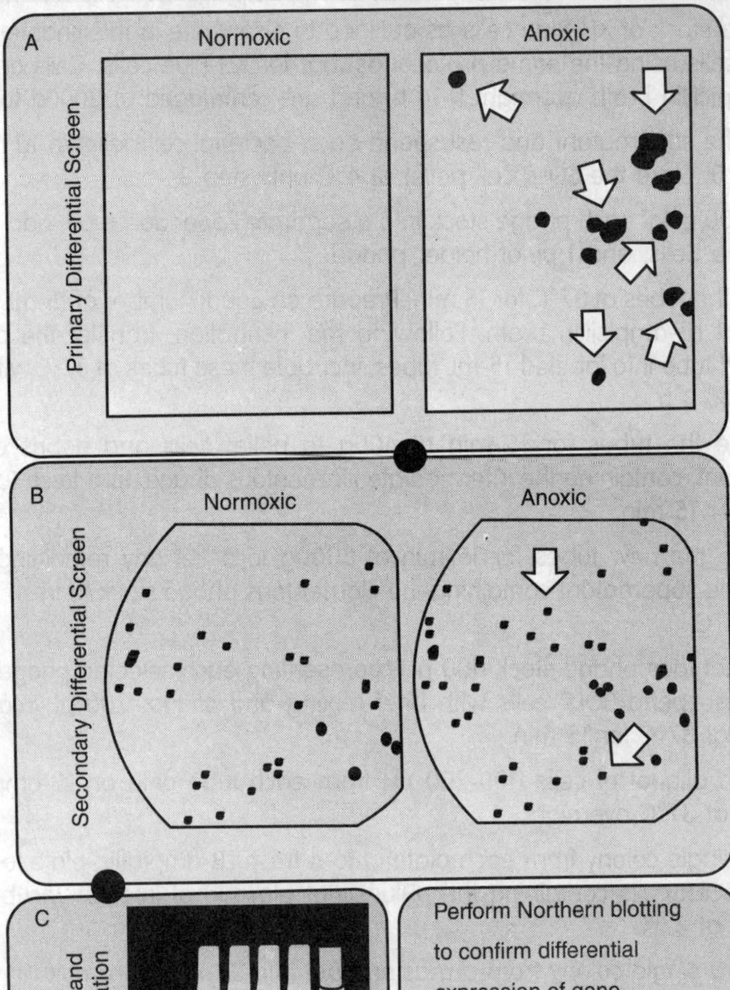

Figure 18.1: Differential screening of a cDNA library. (A) Primary differential screen of duplicate plaque lift membranes using [^{32}P]dCTP labelled cDNA synthesized from mRNA isolated from the hepatopancreas of the marine snail, Littorina littorea. *Normoxic: mRNA isolated from animals kept at 4°C under normal aerated conditions. Anoxic: mRNA isolated from animals kept under a nitrogen gas atmosphere for 1 h, 12 h, and 24 h (equal amounts of mRNA from each time-point combined). (B) Secondary differential screen of duplicate plaque lift membranes using the same probe as in (A) to confirm putative anoxia-responsive clones. (C) In vivo excision of plasmids from phage stock, with subsequent plasmid purification from bacteria, allowed insert isolation from putative anoxia-induced clones.*

5. Add an equal volume of freshly diluted alkaline lysis buffer and keep tube at room temperature for 5 min.
6. Add an equal volume of neutralization buffer and vortex-mix the tube until the white precipitate (protein, genomic DNA, and cell debris) is broken up. Keep on ice for 5 min and then centrifuge in a microcentrifuge at maximum speed for 10 min at 4°C.
7. Transfer the supernatant to a fresh 1.5-mL Eppendorf tube and add an equivalent volume of isopropanol. Invert tube repeatedly to mix and incubate for 10 min at room temperature.
8. Centrifuge the tube at maximum speed in a microcentrifuge for 10 min at room temperature. Aspirate and discard the supernatant and wash the pellet (DNA plasmid) with 70% ethanol. Recentrifuge as described previously for 5 min, Aspirate ethanol and allow the pellet to airdry for 10 min at room temperature.
9. Resuspend the plasmid DNA in an appropriate volume of sterile ddH_2O (generally start with 20 µL) and quantify using spectrophotometry (1 OD = 50 µg/mL of double- stranded DNA). Plasmid is stored at 4°C (short term) or -20°C (long term) until further use.

Introducing Plasmid DNA

1. Digest the plasmid DNA with appropriate restriction enzymes (in this case, *EcoRI* and *XhoII*).
2. Combine an aliquot of plasmid DNA (5 µL) with 0.5 pL of each enzyme, 1 µL of IOx reaction buffer, and adjust the final volume up to 10 µL with sterile ddH_2O. Incubate digest at 37°C for at least I h.
3. Stop the restriction digest by heating at 94°C for 5 min and then adding 0.1 volumes of DNA loading dye.
4. Load samples onto a 1 % TAE agarose gel and separate the restriction fragments by running the gel at 90 V in 1 × TAE buffer. An appropriately sized DNA ladder should be used as a molecular weight standard, allowing the size of the insert to be estimated. An example of digested inserts separated using agarose gel electrophoresis is shown in Figure elsewhere in this chapter.
5. Carefully slice each insert band out of the gel using a sterile razor blade and place each into a labeled Eppendorf tube. To isolate the DNA from the gel agarose, a number of options are available. Commercial kits using filtered columns or glass milk are available, however a quick and easy protocol is outlined here.
6. Place the gel slices at -80°C until frozen (~15 min). The collar of a 1-mL filtered pipet tip is removed using a razor blade and the bottom half of the tip is fixed inside it. This apparatus is then placed firmly into the top of a labeled Eppendorf tube. Each frozen gel slice is transferred quickly into the opening of the tip on top of the filter and the tube is placed into a microcentrifuge set to spin at 10,000g for 3 mnin. Remove the tubes from the centrifuge and discard the filter apparatus. Insert DNA is present in the eluant and can be stored at 4°C (short term) or-20°C (long term).

Synthesis of Probe

1. Transfer an aliquot of the insert DNA (12 µg) into a sterile 1.5-mL Eppendorf tube and

adjust the final volume of 8 ML. Denature the DNA by placing at 94°C for 3 min, centrifuge briefly to collect contents at the bottom of the tube and place immediately on ice. Add the following to the tube on ice: dNTPs (ATP, TTP, GTP; final concentration = 1 mM each), 1 pL random primer (100 mM $d(N)_6$; NEB), 2 µL of lOx random primer buffer, 2.5 U of Klenow fragment of DNA polymerase 1 (NEB), 2.5 µL of [a*-^{32}P]dCTP (3000 Ci/mol; Amersham).

2. Gently mix the tube and place in a heating block at 37°C for 45 min. Load the labeled reaction onto a quick spin column and centrifuge at 400g for 3 min. The eluted radiolabeled probe is used for Northern blotting.

Method of Northern Blotting

1. Prepare a 1.25% agarose formaldehyde denaturing gel and submerse it in enough 1 × MOPS buffer to cover the wells. Pre-run the gel for 15 min (as RNA samples are being prepared).
2. Aliquot the appropriate volume of total RNA, containing between 10 and 20µg, into labeled tubes on ice. Add an equal volume of RNA sample buffer into each tube.
3. Incubate samples at 55°C for 10 min and place immediately on ice. Add the appropriate amount of RNA loading buffer (final concentration = 1 ×) to each sample.
4. Mix the RNA samples gently and briefly centrifuge tubes to collect the entire sample at the bottom of each Eppendorf tube.
5. Load the entire contents of each tube into the wells of the gel and record sample order for reference.
6. Perform gel electrophoresis at 100 V until the loading dye front reaches the end of the gel.
7. Place the gel on plastic wrap and visualize the RNA using a UV light. A photo-graph (digital or Polaroid) should be used to maintain a photographic record. The ribosomal RNA (rRNA) bands are used as an indicator of degradation and as a rough guide to ensure equal loading of the wells. Figure elsewhere in this chapter (lower panel) provides an example of ethidium bromide stained ribosomal RNA visualized with UV light.
8. The separated RNA will be transferred from the gel onto a 0.2-pm nylon membrane by the neutral transfer method. Soak the gel in 10× SSC for 10-15 min to remove excess formaldehyde and ethidiurn bromide.
9. While the gel is soaking, set up the transfer apparatus. Fold a large square of plastic wrap over onto itself (2 ft × 3 ft) and place on the benchtop.
10. Center a square piece of Plexiglas or glass sequencing plate (larger than the gel being transferred) on the plastic wrap. The wrap should extend 3-4 in. beyond the edges of the plate.
11. Cut two pieces of filter paper (used as a wick) such that they are approx. 1-2 inches longer than the plate on all sides.
12. Prewet the plate with 10× SSC and place each filter paper sequentially over the plate. Remove wrinkles or air bubbles by smoothing each layer with a sterile pipet.
13. Orient the gel face down on the second piece of filter paper, directly in the center.

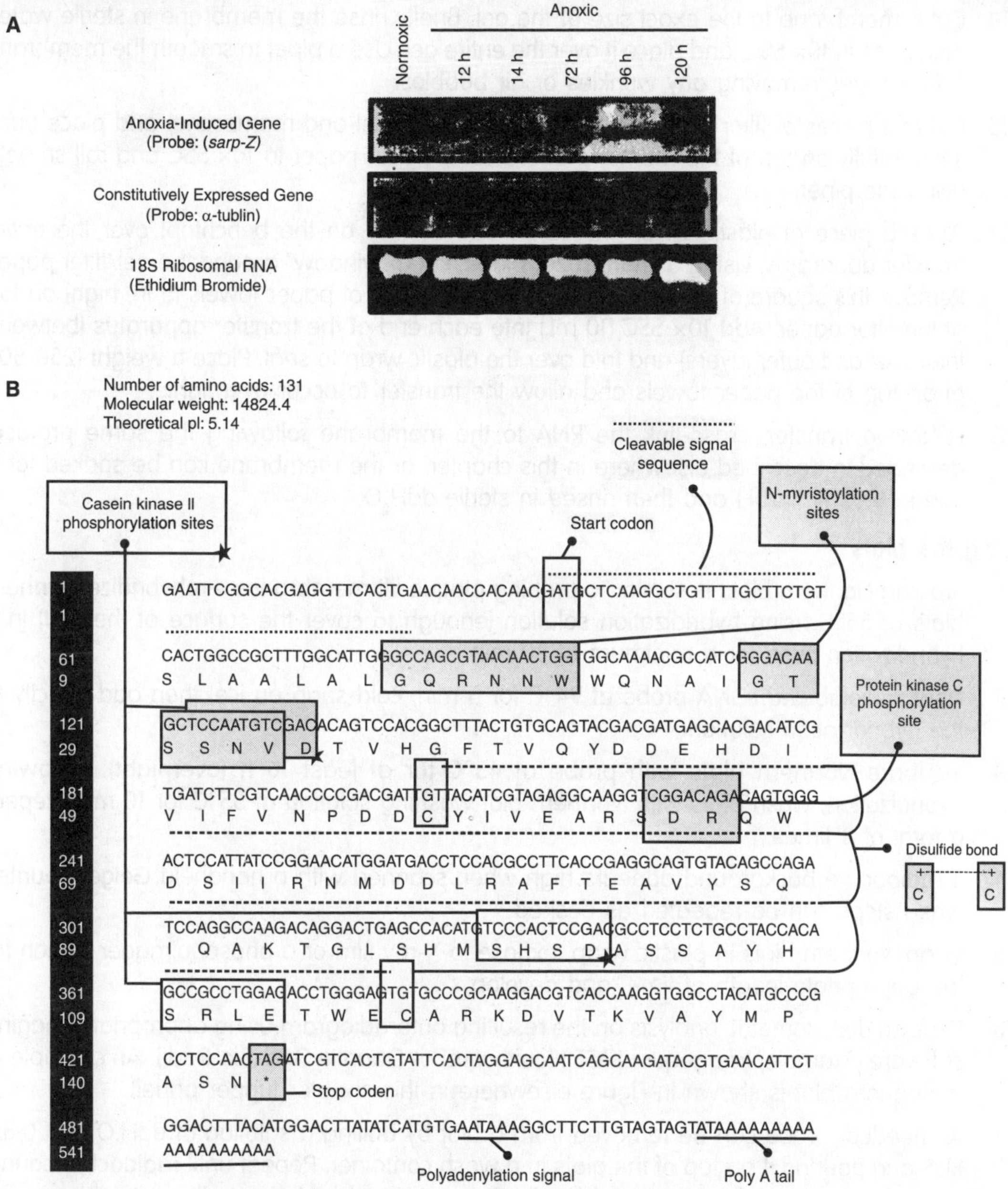

Figure 18.2: Characterization of the novel gene from L. littorea, sarp-2. *(A) Following, insert isolation from a clone designated* carp-2, *expression of the* sarp-2 *transcript was monitored over a time course of anoxia exposure by Northern blotting. The expression of* sarp-2 *was normalized by examining the expression of a control gene (a-tubulin) on the same blot, as well as by examining the intensities of the ribosomal RNA band on the gel before transfer. (B) Sequencing of the insert* sarp-2 *showed a 550 nucleotide sequence containing the full open reading frame for a 131 amino acid protein. Multiple common protein domains and motifs were identified in the* sarp-2 *mRNA and deduced SARP-2 protein sequences using the applications outlined is described elsewhere in this chapter. and selected results are illustrated here.*

14. Cut a membrane to the exact size of the gel. Briefly rinse the membrane in sterile water, then rinse in 10× SSC, and place it over the entire gel. Use a pipet to smooth the membrane over the gel, removing any wrinkles or air bubbles.
15. Cut two pieces of filter paper (the same size as the gel and membrane) and place each sequentially on top of the membrane. Prewet the filter paper in 10× SSC and roll smooth using the pipet.
16. Place a piece of plastic wrap (the same size as that on the benchtop) over the entire transfer apparatus. Using a sharp razor blade, cut a "window" around the gel/filter paper. Remove this square of plastic wrap and place a stack of paper towels (5 in. high) on top of the filter paper. Add 10× SSC (10 mL) into each end of the transfer apparatus (between the inner and outer layers) and fold over the plastic wrap to seal. Place a weight (250-500 g) on top of the paper towels and allow the transfer to occur overnight.
17. Following transfer, cross-link the RNA to the membrane following the same protocol described in described elsewhere in this chapter, or the membrane can be soaked for 5 min in 0.05 N NaOH and then rinsed in sterile ddH_2O.

Probing the blots

1. Probing Northern blots is similar to probing plaque lift membranes. Prehybridize Northern blots at 55°C using hybridization solution (enough to cover the surface of the blot) in a hybridization oven for a minimum of 30 min.
2. Heat radiolabeled cDNA probe at 94°C for 5 min, cold-snap on ice, then add directly to the hybridization solution.
3. Incubate Northern blots with probe at 45°C for at least 16 h (overnight). Following hybridization, wash blots with Northern blot washing solution at 55°C for 10 min. Repeat a total of 4 times.
4. If radioactive background appears high when scanned with a handheld Geiger counter, wash steps can be repeated as desired.
5. Wrap Northern blots in plastic wrap, expose to X-ray film or a phosphoimager screen for the appropriate length of time, and develop.
6. Perform densitometric analysis on the resulting autoradiogram using appropriate imaging software (Example, Imagequant (GE Healthcare) or QuantityOne (Bio-Rad)). An example of a Northern blot is shown in Figure elsewhere in this chapter (upper panel).
7. As needed, probe can be removed from a blot by boiling a solution of ddH,O and 0.5% SDS and pouring it on top of the blots in a wash container. Repeat until radioactive counts reach (or are below) background levels. Rinse stripped blots in sterile water (to remove residual SDS) and re-probe as desired. Reprobing for a constitutively expressed or "housekeeping" gene is used to normalize the relative levels of your differentially expressed gene against a gene whose expression does not change under the experimental condition being examined. The Northern blot in Figure elsewhere in this chapter was stripped and reprobed with a-tubulin to control for loading, as illustrated in the Figure elsewhere in this chapter (middle panel).

8. Additional experimental protocols can and should be performed to further charac-terize a novel gene. The promoter of the gene can be examined to determine if stress or condition responsive elements are present which may play a role in gene expression. Transcriptional analysis can be examined with nuclear runoff assays to determine if a gene is transcriptionally regulated or if the mRNA is stabilized. The gene can be expressed as a protein and basic biochemistry can be carried out to determine a possible role or function *in vitro*. The protein can also be used as an antigen for producing an antibody that can be used for Western blotting and subcellular localization experiments. A knockdown (using siRNA) or knockout (using genetic manipulation) of a gene can be performed to determine the effects and/or phenotype resulting from loss of the novel gene. These methods and techniques are outside the scope of this chapter and they represent downstream experiments from the "starting point" outlined here. Refined protocols can be found in other volumes of the *Methods in Molecular Biology* series, as well as in classic laboratory manuals such as *Molecular Cloning: A Laboratory Manual*. Figure elsewhere in this chapter provides a basic view of experiments used to examine the novel gene, *kvn,* discovered and characterized by the authors.

Analyzing Clones

Differentially expressed clones should be bidirectionally sequenced using standard methods (either *in* the laboratory *or* by a DNA/Genomics facility). Subsequent sequence analysis is performed using commercially available computer software and a wide array of analysis tools available online.

An index of some excellent bioinformatics programs and applications available to the scientific public for analysis of cDNA and putative amino acid sequences is listed in this subheading. As a first step, DNA sequences should be copied and pasted into NCBI's BLAST server for comparison to existing and annotated sequences. Genes sharing sequence identity or similarity to existing expressed sequence tags (ESTs) will be identified with an E-value.

The E-value represents the proba-bility of each alignment occurring due to chance, where a value $<e^{-5}$ signifies that an alignment is not due to error. Novel or unknown genes will often come back with no determinable match, although some conserved domains and/or motifs may be identified and provide a short sequence match.

1. ***Basic Local Alignment Search Tool (BLAST):*** http://www.ncbi.nlm.nih.gov/ BLAST/ BLAST first lists four letter "words" of residues in one sequence; then the words are expanded to segments (profile based). The programs used for nucleotides are blastx (compares a nucleotide query sequence translated in all reading frames against a protein sequence database) and tblastx (compares the six-frame trans-lations of a nucleotide query sequence against the six-frame translations of a nucleotide sequence database), whereas blastp (compares an amino acid query sequence against a protein sequence database) is used for protein analysis.

2. ***ORF (open reading frame) finder:*** http://www.ncbi.nln.nih.gov/gorf/gorf.html ORF finder allows identification of potential coding regions in all 6 reading frames of a nucleotide sequence. Selected ORFs can then be transferred directly into BLAST to search for conserved domains and similar sequences using blastp (for proteins).

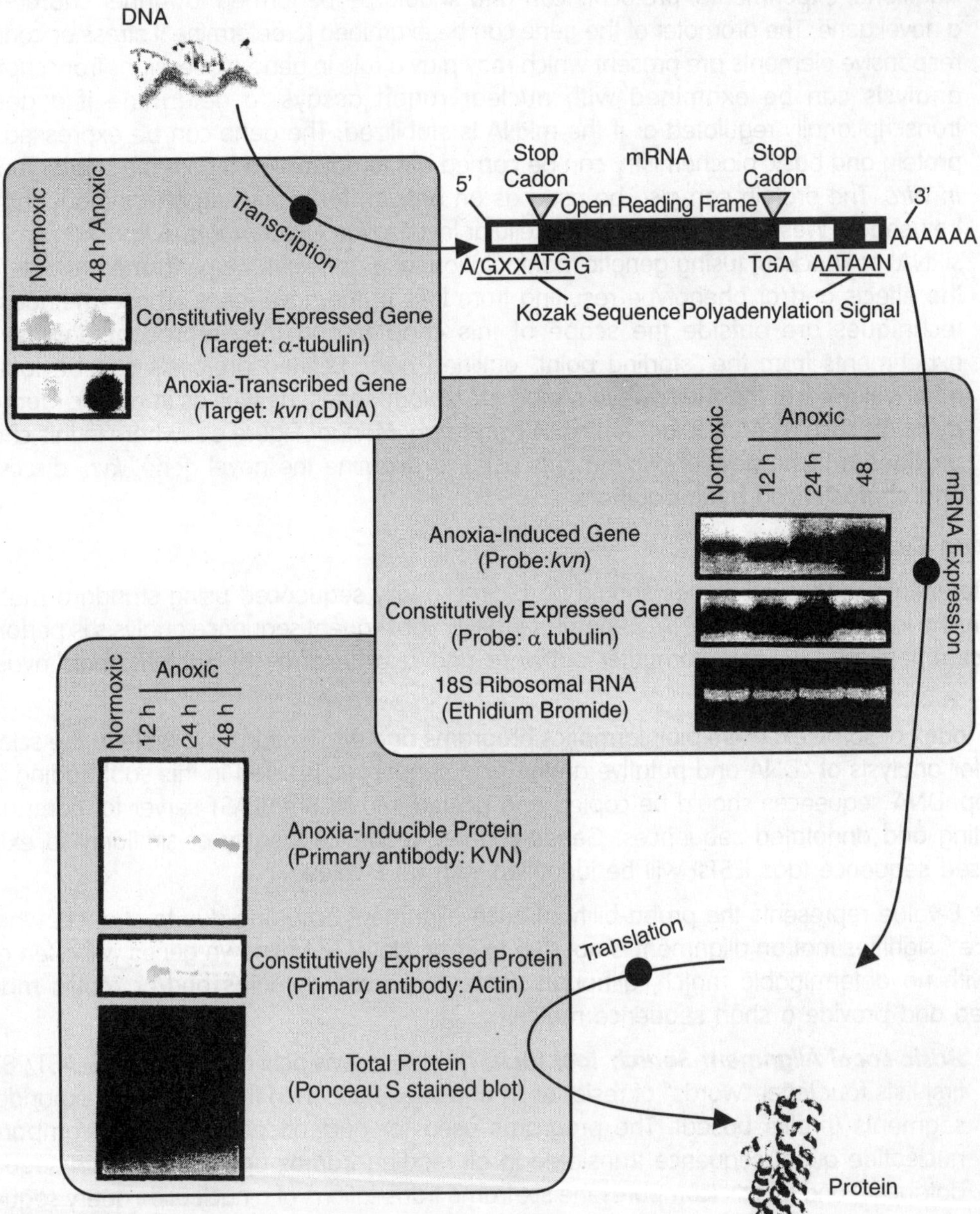

Figure 18.3: Flowchart outlining the characterization of a novel gene. Starting with genomic DNA, the transcription of a gene can be measured using a nuclear run-off assay (top left panel). The gene examined here (kvn) is actively transcribed in nuclei isolated from the hepatopancreas of L. littorea *kept under anoxic condi-tions for 48 h. This gene has been characterized previously (4) and contains all of the features common to eukaryotic genes that are identifiable in mRNA. Expression of kvfn was monitored over a time course of anoxia exposure as shown by Northern blotting (middle panel).*

Translation and Determination of a Protein Sequence

Determining the function of novel genes is a challenge currently facing many of the large genome projects. An mRNA sequence contains much infor-mation, but it is traditionally the functioning protein that performs the important task.

In order to anticipate the function of a particular gene, its sequence is usually translated into a predicted protein, which is then analyzed for structural homology and conserved regions (known as domains or motifs) that have been identified in proteins having known functions. Many times a cDNA clone will represent a known homologue.

The genes/proteins represented by these clones can be assimilated into an existing scheme, since they have often been charac-terized in other systems, organs or tissues (as well as during various stresses or conditions) and perform a known role or function. However, if the clone of interest does not have any known homologues, it must be further analyzed and characterized using the tools available. Some tools for analyzing a novel gene/protein include the following:

PredictProtein:

http://www.embl-heidelberg.de/predictprotein/predictprotein.html

The translated protein sequence of interest is queried against characterized sequences using the PredictProtein server, which retrieves similar sequences in the database and predicts aspects of protein structure.

Protein Sequence Analysis (PSA):

http://bmerc-www.bu.edu/psa/request.htm PSA predicts protein secondary and tertiary structure based on amino acid sequence.

SOSUI:

http://sosui.proteome.bio.tuat.ac.jp/sosui submit.html

This tool predicts secondary structure of membrane proteins from an input protein sequence by examining the physicochemical properties (such as hydrophobicity and charges) of amino acid sequences.

PROSITE:

http://www.expasy.ch/prosite/

PROSITE is a database of conserved protein motifs. The query output provides matches for biologically significant patterns and profiles based on the primary sequence of the protein.

PSORT:

http://psort.nibb.ac.jp/

After a gene is translated into a protein, the amino acid sequence often contains a region that represents a "transport" or "localization" signal. This signal determines the cellular localization of the protein and may help in determining its function. PSORT compares an input sequence to known consensus sequences to determine if a sorting signal is present.

SignalP:

http://www.cbs.dtu.dk/services/SignalP/

The SignalP server predicts the presence and location of signal peptide cleavage sites in input amino acid sequences. Submitting DNA Sequences to NCBI GenBank

http://www.ncbi.nlm.nih.gov/Genbank/submit.html

Prior to publication of research that has retrieved and characterized DNA sequences, the sequence must be annotated and submitted to a sequence database. The submission may include such information as coding regions, start and stop codons, or any information provided by the applications listed here. Using an internet browser, the new sequence is submitted to Genbank through Bank It (web based tool) or Sequin (stand alone submission tool).

POINTS TO REMEMBER

1. Total RNA sample quality can also be assessed via denaturing agarose gels. Both the integrity of the ribosomal RNA (rRNA) and the distri-bution of mRNA in each lane are indicators of degradation. The 28S rRNA band should be approximately twice as intense as the 18S rRNA (although with many inver-tebrates, only a single rRNA band is observed due to a cryptic nick in the 28S rRNA, resulting in a single rRNA band migrating at the approximate size of the 18S rRNA).

   There should be an even distribution of mRNA throughout the lane (as indicated by a light pink ethidium bromide smear when viewed using UV light).

2. The final reaction volume in the cDNA library synthesis step will be 50 µL, so the RNA must be resuspended in the appropriate volume of ddH_2O. If the RNA is not concentrated enough, the sample can be concentrated using a SpeedVac centrifuge (preferred method) or reprecipitated using isopropanol as before (some loss will occur).

3. Remove 5 µL of the first-strand synthesis reaction and aliquot it into a new tube. Add 0.25 µL of [α-^{32}P]dCTP and incubate with the first-strand synthesis reactions.

   Place the radioactive first-strand synthesis control reaction at -20°C until ready to resolve by electrophoresis on an alkaline agarose gel.

4. Remove an aliquot of the second strand cDNA synthesis reaction (1 µL) and analyze it with the aliquot saved from the first strand reaction using an alkaline agarose gel. These radiolabeled "control" reactions allow determination of the size and quality of synthesized cDNA. Gel electrophoresis is performed similar to the procedure outlined in described elsewhere in this cahpter Melt agarose (3.75 g) in ddH_2O (270 mL) containing ethidium bromide (1 pg/mL) and cool to 60°C. Add 0.1 volumes of lOx alkaline agarose electrophoresis buffer (500 mM NaOH, 10 mM EDTA, pH 8.0), swirl without introducing bubbles, pour gel, and allow it to polymerize. Submerse the gel in enough lx alkaline agarose electrophoresis buffer to cover the wells and prerun for 5 min. To each of the samples (first- and second-strand reactions, 1 µL), add 10 µL of I × alkaline agarose electrophoresis buffer and 2 µL of alkaline loading buffer (300 mM NaOH, 6 mM EDTA, 18% Ficoll 400, 0.15% bromocresol green, 0.25% xylene cyanol FF). If possible, electrophorese gel in a cold room (or at low voltage) to prevent overheating. Following electrophoresis, fix the gel in 7% trichloroacetic acid (w/v), dry in a gel dryer, and perform autoradiography.

5. To enhance precipitation, add NaCl to a final concentration of 25 mm.

6. Depending on how mRNA was originally prepared, the quality may be suitable to create a library without size fractionation. If transcripts above or below a certain size are desired, cDNA fractionation can be performed. One method for this is briefly outlined here. A Sepharose drip column (~1 mL) can be used to separate cDNA by size; longer cDNA sequences will elute off the column first followed by progressively shorter sequences. Columns can be made in the lab using Sepharose having the desired molecular weight cutoff (often S500), or can be purchased from a company such as Invitrogen). Equilibrate column with loading buffer (10 mM Tris-HCl, pH 7.5, 0.1 mM EDTA, 25 mM NaCl) by loading l mL onto the column and allowing it to drain completely. Repeat 3-5 times. Load sample cDNA onto the column (in a volume of ~100 μL of loading buffer) and allow it to enter the Sepharose. Aliquot 100 μL of loading buffer onto the column, allow it to enter the Sepharose and collect the eluant into the first tube. Continue adding 100-μL aliquots of loading buffer on top of the column (allow each to enter the Sepharose prior to adding subsequent aliquots). Collect the eluant in approx. 50 μL fractions. A few microliters of a sample dye containing glycerol (50%) and bromophenol blue (50%) can be added after the sample is loaded to track the progress of the cDNA elution. Each collected fraction can be analyzed by loading 5 pL on a 5-6% standard nondenaturing acrylamide gel, followed by autoradiography (wet or dry gel), to determine cDNA size in each. Combine the fractions containing the desired lengths of cDNA into a single tube and then process as described elsewhere in this chapter. Resuspend the dried pellet in the appropriate volume of ddH_2O and proceed to elsewhere in this chapter.
7. The Uni-ZAP vector contains β-galactosidase within the multiple cloning site, allowing a "blue-white" color assay to be performed to determine the ratio of recombinants (white plaques) to nonrecombinants (blue plaques) in the cDNA library. Transfer an aliquot of the final packaged reaction (1 μL) into two Eppendorf tubes. Label the first tube 1 (undiluted). Add 9 μL of ddH_2O to the second tube, then aliquot a single microliter of this 1:10 dilution into a fresh Eppendorf tube labeled 2 (1:10 dilution). Add an aliquot of XL 1-Blue cells (100 μL; prepared as describe elsewhere in this chapter) into each of the labelled tubes and incubate at 37°C for 15 min. Combine IPTG (15 μL of a 0.5 M stock) and X-gal (50 μL of a 250 mg/mL stock) with 3 mL of molten NZY top agar. Cool the agar to <50°C, add the infected XLI-Blue cells, spread the mixture on top of a standard NZY plate, and allow top agar to solidify. Incubate the plate at 37°C overnight (~ 16 h) until colored plaques develop. Examine both plates (one of which should contain 250-500 plaques) and count blue vs. white plaques to determine the background of the library.
8. Sheared salmon sperm DNA (Sigma-Aldrich) can also be included as a non-specific, unlabelled DNA source to reduce background binding to membranes. Add sheared salmon sperm DNA (0.2 mg DNA/mL hybridization solution) to the Eppendorf tube containing probe and follow the protocol from elsewhere in this chapter. This note is also applicable to described elsewhere in this cahpter.
9. When adding probe to the hybridization tube, take care not to "spray" the membrane. If concentrated probe contacts the membrane, the result can be a "hot spot" which may obscure important regions of the blot. Probe should be dispensed directly into hybridization solution (using longer pipet tips) or added to an aliquot of hybridization solution (~1mL),

which is carefully transferred to the tube drop by drop.

10. A layer of plastic wrap can be placed around the stack to prevent a potential "leak" (and resulting short circuit) in the transfer setup between the paper towel and the wicks. This extra layer ensures that buffer must flow through the gel and membrane to reach the paper towel, improving Northern blot transfer.

Chapter 19

Determining Changes in Gene Expression

Complementary DNA (cDNA) *macroarrays* are useful for detecting differ-ences in gene expression caused by exposure to *environmental toxicants*. In this procedure, cDNA clones of interest are selected and immobilized on nylon membranes.

RNA is isolated from a selected tissue, reverse *transcribed* to cDNA in the presence of radioactive nucleotides, and then *incubated* with the prepared *nylon membranes* to allow the labeled cDNA to bind to its complement on the array. Intensity of the *signal* is proportional to the amount of message in the sample RNA pool. For many investigators, cDNA arrays have become useful tools to identify potential new biomarkers. By using a few well-designed arrays, *hypotheses* can be *generated* and tested to gain greater insight into the direct or indirect global effects of *toxicants*, an approach that has been termed *"Systems Toxicology"*.

Often, a *toxicant* may exert influence on *biochemical pathways* not previ-ously considered. Through the use of gene arrays, one can get a better under-standing for the mechanisms by which *toxicants* have genome-wide effects. In this way, small-scale theories can be moved into a larger-scale view to allow for a better understanding of the mechanisms by which toxicants work. Gene arrays offer the *investigator* the *ability* to screen for many different responses *simultaneously* from a single RNA sample.

What previously would have taken larger amounts of RNA and longer time periods can be done in a fraction of the time using this technology. eDNA arrays, like other techniques, have their own advantages and disadvantages. The greatest advantage is the ability to *simultaneously* screen a large number of genes for changes in expression.

Another advantage is that the starting sample size can be much smaller (just two micrograms of total sample RNA per array) compared to the 10-30 µg required per lane for *Northern blotting*. An important disadvantage to this technology, however, is the inability to determine different *transcript* sizes or to locate *transcriptional* start sites, which is more easily accomplished using Northern blotting or RPA (RNase protection assay). RPA uses solution *hybridization*, which has the advantage of being *quantitative* and fast, but requires synthesis of a labeled probe that can sometimes be *troublesome* to synthesize.

In today's world of modern molecular biology, we have seen the advent of *numerous* genome projects allowing for gene arrays to be made based on *putative* genes. By doing this, companies have created microarrays that can theoretically contain all gene *transcripts*.

Toxicological research examining gene expression within traditional models (like *human cell lines, rodents*, and *zebrafish*) has become easier and more efficient than for those using *nontraditional* species, such as most fish species and *invertebrates*.

For those who use nontraditional species, species-specific arrays need to be created using cDNA clones already on hand. These are the people for whom this chapter is written. People who are studying species that are *nontraditional* cannot simply order clones off the shelf, like those that are available for human and mouse. For species where gene clones are not *commercially* available, libraries of genes must be generated in house. One method gaining in *popularity* is *suppressive-subtractive* hybridization from *Clontech*.

This procedure allows for "*subtraction*" of one pool of cDNA from another (these can be two treat-ments or a treatment from control). A second, older method is differential display (GenHunter Corporation, Nashville, TN is one supplier) that allows for direct selection of differentially expressed genes from a gel for *cloning* and *identification* by sequencing. While still a useful method, it has fallen out of *regular* use in the past few years.

REQUIREMENTS

Obtaining Clones

1. Glycerol stocks of *E. coli* containing *plasmids* with cDNA inserts of' interest.
2. Taq *polymerise* and buffer (New England Biolabs cat. no. M0267).
3. 10 m*M* dNTP mix (any commercial source).
4. Primers to amplify the insert in your cloning vector diluted to 10 μM (M13, SP6, T4 or T7 are common ones).
5. 70% Ethanol in nuclease-free water.
6. Tris-EDTA buffer: 10 mM Tris, 1 mM EDTA, pH 8.0.
7. 96-Well culture plates (any commercial source, round-bottom plates are useful for overnight cultures).
8. Breathe easy strips for 96-well plates (USA Scientific cat. no. 9123-6100).
9. LB medium: 10 g of tryptone, 5 g of yeast extract, 10 g of NaC1/L, pH 7.0.
10. Millipore *filtration* system for purifying the *polymerase* chain reaction (PCR) products (MultiScreen Filtration System Vacuum Manifold cat. no. MAVM0960R and Montage PCR_{96} Cleanup plates cat. no. MANU03010).
11. 96-Well PCR plates that fit your *thermocycler* (any *commercially* available plates for PCR will serve the purpose).
12. Multichannel pipet, 20-μL and 200-μL sizes will be very helpful.
13. Flat bottom 96-well or 384-well plates (any commercial source).

14. Agarose gel electrophoresis set up capable of running the PCR products.
15. DNA molecular weight ladder that has size range from 300 base pairs to 2000 base pairs (from any commercial source).
16. UV spectrophotometer, preferably a plate reader model.
17. 96-Well plates capable of being read in a UV spectrophotometer if desired (any commercially available, check your machine for suggestions).

Method of Printing

1. Precut Pall Biodyne B neutral nylon membranes (Nunc, cat. no. 250385).
2. Spot Report Array *Validation* System (Stratagene, cat. no.252005-7) which contains *several* controls [poly(dA), human Cot-I, *Arabodopsis* cDNAs, empty cloning vector, etc.] which should be spotted on the array.
3. *UV-transilluminator* (we use a UV Stratalinker 1800, *Stratagene*).
4. Centrifuge with a rotor for holding 96-well plates (any model will serve the purpose).
5. Robot (i.e., a Biomek 2000, Beckman Coulter using 100-nL pins).
6. Distilled water.
7. 10% Bleach.
8. 70% Ethanol.
9. Bromophenol blue.
10. 20× SSC: 3 *M* NaCl, 0.3 *M* sodium citrate, pH 7.6, in nuclease-free sterile water. *Autoclave* to sterilize and store at room temperature. 0.01 m*M* *bromophenol* blue is added after *sterilization.*

Hybridization

1. M-MuLV reverse transcriptase enzyme and buffer [New England Biolabs, cat. no. M0253S (200 U/ml)].
2. Random hexamer primers (New England Biolabs, cat. no. S 1254S) add 80 µL of nuclease-free sterile water to 1 A_{260} unit.
3.10× dNTP mix: can be purchased commercially or can be made by combining 20 µL each of 100 m*M* dCTP, dGTP, dTTP, and 10 µL 100 mill dATP. Bring the volume to 400 µL by adding nuclease-free sterile water (gives a final *concentration* of 4 m*M* dCTP, dGTP, dTTP, and 2 m*M* dATP).
4. [α-^{32}P]dATP: from any supplier (Perkin Elmer cat. no. NEG612H for 250 µCi).
5. Spike RNA mix: commercially available from Stratagene as a kit with control spikes for printing the arrays or individual spikes can be purchased (Spike 2 RCA and Spike 3 rbcL, cat. no. 252202 and 252203, respectively).
6. 20× SSC: 3 *M* NaCl, 0.3 *M* sodium citrate, pH 7.6, in nuclease-free sterile water. Autoclave to sterilize and store at room temperature.

7. 20% SDS: 200 g of sodium dodecyl sulfate in 1 L of sterile water.
8. Hybridization buffer: 0.375 *M* NaCl, 0.0375 *M* sodium citrate, 7% SDS and 25% fresh *formamide* in a final volume of 500 ml- made up with *sterile* nuclease-free water. Add 100 mg of yeast tRNA. Store at 4°C.
9. Wash buffer 1: 2× SSC-0.5% SDS (100 rnL of 20x SSC, 25 mL of 20% SDS, and 875 mL of water).
10. Wash buffer 2: 0.5× SSC-0.5% SDS (25 mL of 20× SSC, 25 mL of 20% SDS, and 950 mL of water).
11. 10 mM EDTA, pH 8.0.
12. Kit to remove excess nucleotides from the labeling reaction (like Qiagen QIAquick Nucleotide Removal Kit, cat. no. 28304).
13. Dry bath incubators.
14. Liquid scintillation counter and buffer.
15. Hybridization bottles (Fisher Scientific).
16. Hybridization oven (we use a MaxiOven from Labnet).

Method of Scanning

1. Transparencies: from any office supply store.
2. Phosphor screen: Molecular Devices.
3. Phosphor imager: We use Molecular Devices Typhoon Scanner.
4. ImageQuant software: Molecular Devices.

PROCEDURES

Process of Amplification

1. Dispense 98 µL of LB broth containing appropriate *antibiotics* into each well of a 96-well plate.
2. Add 2 µL of each glycerol stock containing the *plasmid* of interest to the LB (one clone per well). Also, include an empty *vector* as another control for your array (should have very low *hybridization* when hybridized with your sample).
3. Cover the plate with a breathe easy strip and replace lid on the plate.
4. Allow the cultures to grow overnight at 37°C with shaking (about 125 rpm).
5. The next morning, make up a PCR master mix for 90 µL per reaction (make enough for 100 reactions if using a full plate). See Figure elsewhere in this chapter for a schematic for this method.
6. *Dispense* 90 µL of PCR master mix to each reaction well, and add 10 µL of overnight bacterial culture to each well.
7. Cover the PCR plate with an appropriate cover and cycle as described in Table elsewhere

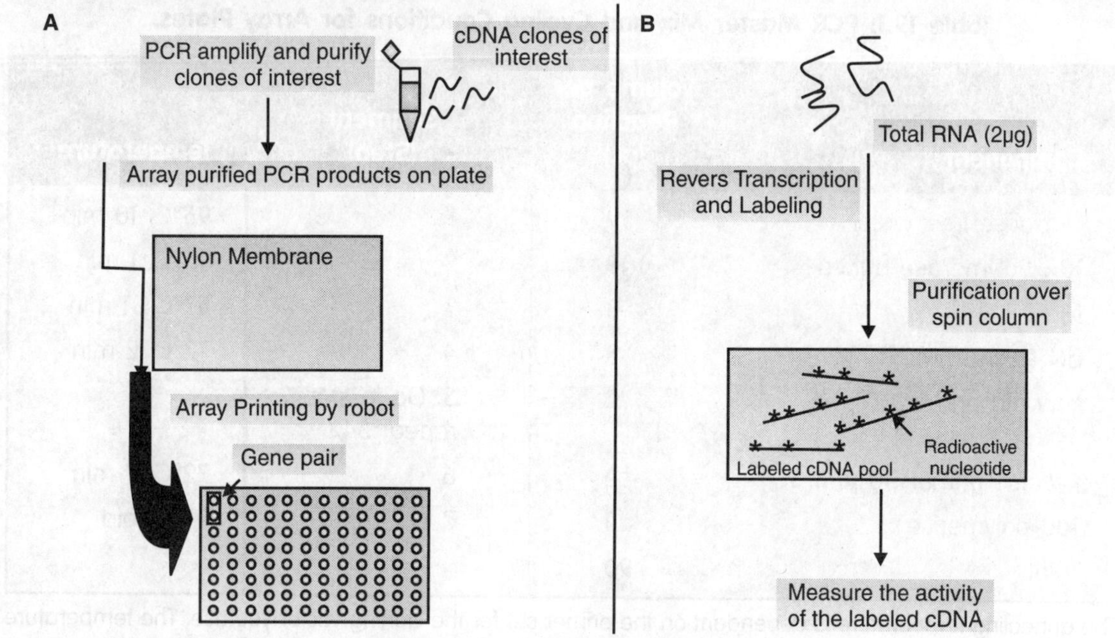

Figure 19.1: Schematic of array printing process and cDNA labeling reaction. (A) General flowchart for the major steps involved in printing a nylon cDNA gene array. Starting from glycerol stocks containing the genes of interest, one can print an array. (B) The major steps for starting from total RNA to having labeled single-stranded cDNA.

in this chapter.

8. After cycling, analyze the PCR products on a 1 % agarose gel to ensure that clones were amplified with 5 μL of each product along with a DNA ladder. Re-run the PCR for any clones that did not properly amplify.
9. After you know that the clones have all amplified, load them onto the Millipore filtration unit and allow the fluid to flow through the membrane. Wash once with 70% ethanol.
10. Turn off the vacuum and resuspend the product by adding 50 μL of TE buffer, and incubate at room temperature for several minutes, and then pipet (a multichannel pipet is useful for this) into another 96-well plate.
11. Measure the concentration using a UV-spectrophotometer by diluting 5 μL of purified DNA in 95 μL of water.
12. If necessary, reamplify any clones that did not provide enough material and these can be pooled with the first round if desired. You will need at least 160 ng/μL for printing.

Mechanism of Printing

1. Before starting this step, you should plan the layout of genes on your array. We always spot genes in duplicate, keeping in mind that you need to include controls on your array.
2. Dilute purified DNA (3000 ng in 18.75 μL = 160 ng/μL) and place into the desired well for printing.

Table 19.1: PCR Master Mix and Cycling Conditions for Array Plates.

Component	Volume per Reaction (μL)	Segment (Step)	Temperature
Nuclease-free water	59	1	95°C, 10 min
10× Polymerase buffer	10	2	95°C, 1 min
$MgCl_2$	8	3	57°C,[a] 1 min
dNTPs (10 MM)	8	4	72°C, 2 min
Forward primer (10 μM)	2	5: Go to step 2, repeat 34X	
Reverse primer (10 μM)	2	6	72°C, 7 min
Taq polymerise	1	7	4°C, hold
Total	90		

[a]The annealing temperature is dependent on the primer set for the cloning vector you use. The temperature listed is for the M13 primer set.

3. Add 1.9 μL of 3 *M* NaOH to each well.
4. Cover the plate and incubate at 65°C for 15 min (we use the hybridization oven for this).
5. Immediately chill on ice for 2-3 min.
6. Centrifuge the plate briefly.
7. Remove the plate cover and add 9.35 μL of 20× SSC that contains *bromophenol* blue to each of the *wells, pipet up* and *down* (volume should now be ~30 μL).
8. Centrifuge the plate again briefly at lOOg for 5 sec. (It is very important to do this.)
9. Set the *plate, membranes, bleach, water,* and *ethanol* onto the robot and begin printing the arrays. You will need to program your robot accordingly.
10. In between printing sets (when all of the arrays on the platform have been printed) it is necessary to clean the pins. Have the robot *submerge* the pins for 10 seconds each in bleach, then *water,* then *ethanol,* and then use the fan to dry the pins and then print again. While the pins are being cleaned, a new set of membranes can be made ready.
11. Crosslink the DNA to the membrane using a UV-transilluminator at the setting used for Southern blots (100 mJ).
12. Using 100-nL pins with the robot it is possible to print over 100 membranes with high *fidelity*.

Labeling Samples

1. Remove radiation from freezer and keep behind a protective Plexiglas shield. Turn on the 37°C and 100°C water baths.

2. Turn on the hybridization oven and set it to 64°C and place the hybridization buffer in the oven to preheat.
3. In a microcentrifuge tube add 2 µL of random primers, 0.6 µL; of spike RNA mix, 2 µg of sample RNA and add DEPC water to bring the volume up to 13 µL.
4. Place the tube at 64°C for 5 min, and then let it cool down at room temperature for 5-10 min to allow the primers to anneal.
5. After the tubes have cooled, add 2 µL of 10× RT buffer, 2 µL of 10× dNTP mix, l µL of *M*-MuLV enzyme, and 2 µL of [a*-^{33}P]dATP. (Total volume in the tube = 20 ML).
6. Incubate at 37°C for 1.5-2 h.
7. Centrifuge the tubes down before opening.
8. Purify the labeled samples using a nucleotide removal kit.
9. Check counts of the sample on a liquid scintillation counter (we count 2 µL of the labeled material).

Washing

1. Start prehybridizing the membranes. Place the membranes in the hybridization bottles, add 5-6 mL of hybridization buffer, and return the hottlc~ back into in the hybridization oven and rotate (12-14 rpm) at 64°C for 1.5-2 h.
2. Calculate the required volume of *radiolabeled* cDNA to be used for hybridization. Each membrane needs to be *hybridized* in the same concentration of *radioactivity* (I million cpm/mL).

   Formula 1 : Calculation for diluting labeled cDNA

   Probe= 1,000,000 cpm/mL* V/cpm/µL

   where Probe = the number of milliliters of label required in µL and V = volume of hybridization buffer in mL. Then add 10 m*M* EDTA. The amount of EDTA required is 20 times the volume of Probe calculated above.

   Formula 2 : Calculation for the addition of EDTA

   Probe*20 = µL 10m*M* EDTA to be added to the label.
3. Add the volume of probe calculated from formula 1 to volume of 10 MM EDTA computed in formula 2. Combine these in either a 1.5-mL *centrifuge* tube that you can put a locking cap on or into a screw cap tube to prevent the tube top from opening during the *denaturation* step.
4. Denature probe at 100°C for 5 min and then place it on ice for 2 min.
5. Centrifuge the tubes at 2,700 × g for 1 min and add the contents directly into the buffer of the appropriate hybridization bottle, already containing the membrane.
6. Shake the bottle slightly and place it back into the hybridization oven at 64°C. Hybridize overnight, about 14 h at 12-15 rpm.
7. Place wash buffers 1 and 2 in the oven at 64°C (overnight). There is usually enough room

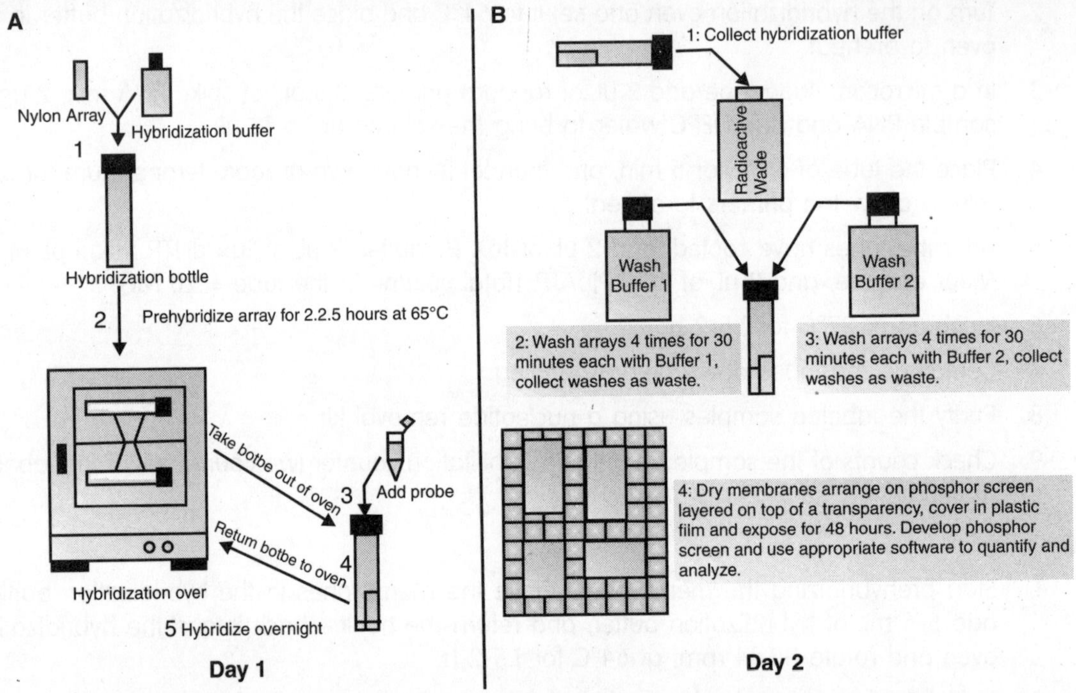

Figure 19.2: Schematic for hybridization and washing of the arrays. (A) The major steps for the hybridization procedure. (B) Procedure for how to wash the arrays the day after hybridization, and laying the membranes out on a phosphor screen. Note the asymmetrical layout of the membranes to facilitate identification after scanning.

on the sides of the hybridization oven to *accommodate* the wash buffers.

8. The following morning you may begin washing the membranes. Instructions for each wash:
   a. First, pour off the hybridization buffer into radioactive waste as required by your institution, then add a couple of inches of wash buffer 1 to each hybridization bottle. Make sure that the buffer covers at least half of the membrane. Return to the hybridization oven for 30 min rotating at 12-14 rpm.
   b. Repeat the wash with buffer I three more times and then wash 4 times using wash buffer 2 (also 30 min each). We collect all washes as *radioactive* waste to be safe, but follow the *guidelines* at your site.
   c. You should complete a total of eight wash steps.
9. Remove the membranes from the bottles, dry them, and then place them in the *phosphor* imager cassette. (Note: By placing the transparency sheet on the grid portion of the cassette and placing the membranes in alignment to the grid, it makes the membranes easier to analyze; *see* Note elsewhere in this chapter) Develop the image of the membrane on the *phosphor* screen for 48 h and scan it with a *typhoon* scanner.

Quantification

After exposure to the phosphor screen, the membranes can be removed from the cassette and the screen can be developed. For spot quantification, we have used the ImageQuant 5.1 software package (Molecular Devices). First, the image is adjusted for contrast (this is only for the eye and does not affect the numbers in later analysis) and then a pair of boxes is drawn around the largest or most intense pair of spots (single gene). This pair of boxes is then copied to the other genes so that the area is *maintained* for all of the *analytes*.

1. After the boxes are drawn and all of the data from all of the membranes are collected, compute the mean of each *gene pair* and subtract the mean of all of the *blank spots* that were designed into the array.
2. Next, the genes on each membrane need to be *normalized* in such a way that they can be compared to other membranes of the set. This may include scaling each membrane to a *particular* mean or median, usually of the highest *membrane*.
3. Next log transform the data set to make the data appear more normally distributed. A *statistician* or *bioinformatics* expert may be helpful for this.
4. Use the appropriate statistical model and method for making your *comparisons* that you have determined prior to running the array experiment.
5. After determining which genes are most interesting, they should be confirmed by other methods, such as real-time or quantitative PCR.

Mathematical Calculation

The most important thing to consider before doing gene array experiments is to decide how to analyze the *final data*. This includes the number of *individuals* per treatment, whether to pool individuals, what *comparisons* are going to be made, and the statistical tests to be used to make these comparisons. Since the methods for analyzing array experiments are still evolving, it would be best to consult with a statistician or *bioinformatics* specialist for assistance before starting these experiments, rather than *afterwards*. There are several ways to depict the data as well and usually this depends on the type of *experiment* that was done.

POINTS TO REMEMBER

1. We always spot the genes in duplicate in order to safeguard against mis-spotting that may occur with the *robot*.
2. When designing the layout of your array, it is important to keep in mind that you need to include *controls*. We use an empty *vector amplicon*, several spots with no DNA (only 20× SSC with bromophenol blue), as well as messages for *ribosomal proteins*, in addition to the *spikes* and other cDNAs that come in the *array spot* reporter kit from *Stratagene*.
3. Since printing the arrays can be costly, expense may be controlled if a targeted array is designed such that two or more arrays are printed on a single membrane.
4. A robot is not absolutely necessary as hand printing stamps are *commercially* available, but to print in this fashion would be difficult for large *quantities* of arrays and they would be less consistent. It may be possible also to use a vacuum driven dot blotter, but this has not been *tried* and would likely be very *tedious*.

5. We generally use 5 mL of hybridization buffer, but the volume used depends on the size of the *hybridization* bottles. The larger the bottle the more buffer you need, as long as half of the membrane remains in contact with the buffer.
6. If you do not have a phosphor imager available to you, you can also use regular X-ray film. However, the *dynamic* range of film is several logs less than that of the *phosphor* screen so resolving differences may become more difficult. If film is used, *photodensitometry* can be used in place of ImageQuant or other *phosphor* imaging software.
7. Placing the membranes in an *asymmetrical* pattern in the phosphor screen will also enable easier identification of which *individual* membrane is which when you develop the phosphor image.

Chapter 20

Alignment with Software Program

With a growing number of partially or completely sequenced genomes, comparative analysis of genomic sequences is becoming a crucial tool for genome analysis and *annotation*. Practically all methods for comparative sequence analysis rely on pairwise or *multiple alignments*, so a*ligning genomic* sequences is the first and most important step for *comparative genomics*. Most computer programs for pairwise or multiple sequence alignment that have been developed during the 1980s or 1990s are either *global* or *local* alignment methods.

Global programs align the input sequence over their entire length, whereas local programs return only the most highly conserved region in the input sequences and ignore the remainder of the sequences. The former methods are based on the famous *dynamic-programming algorit* m proposed by Needleman and Wunsch for global pairwise alignment; extensions of this *algorithm* to multiple *alignment* include CLUSTAL W9, T-COFFEE and many other *approaches*.

Established methods for local alignment include the Smith-Waterman algorithm (and BLAST) for pairwise alignment and GIBBS sampling for local multiple alignment. There are two mar reasons why these *traditional* alignment approaches are of *limited* use for *analyzing* long genomic sequences. First, many sequence families are not related over the full sequence length, but nevertheless exhibit much more sequence similarity than just a single conserved *motif*.

Such sequence families cannot be aligned with traditional global or local alignment methods that would either align the sequences over their entire length or align only one conserved motif. This is the case for syntenic genomic sequences where conserved structures such as protein-coding genes are *interrupted* by *intergenic regions* with little or no sequence *similarity*, but it is also the case for many protein families. In addition, most of the alignment methods developed until the mid-1990s were designed to align rather short sequences, e.g., single genes or proteins.

Consequently, these methods are far too slow for large-scale alignment of genomic sequences. For these reasons, during the last few years, a new generation of alignment program has been developed that can cope with these *challenges*. The program DIALIGN has been developed for multiple alignment of distantly related protein or nucleic-acid sequence families. It combines global and local alignment features by composing pairwise and multiple alignments from local pairwise alignments, so-called fragment alignments or fragments. This way, it is able to detect and align conserved regions in the sequences but *ignores nonrelated* parts of the sequences, but it is not restricted to return the single best local similarity as traditional local alignment programs do.

The idea is that the program itself should decide if or to what extent homology is detectable among the input sequences. Thus, for *globally* related sequence sets, the selected fragments will cover the sequences over their *entire* length and, as a result, the program would return a full *global alignment.*

If only local similarity is *detectable,* however, the program would return a local alignment or an alignment composed of several local homologies separated by *nonhomologous* parts of the sequences. More *precisely,* fragment alignments or fragments are defined as local gap-free pairwise alignments among the input sequences. Each possible fragment is given a quality score based on *statistical significance,* and for a given set of input sequences, the program searches a *collection* of fragments that can be *integrated*

into one resulting (pairwise or multiple) output alignment. This way, DIALIGN tries to align all regions of the sequences that *exhibit* some *statistically significant* degree of similarity-provided they fit into one single output alignment. Because the program is able to integrate an *arbitrary* number of local *alignments* into one output *alignment,* it can deal with many sequence families that cannot be reasonably aligned with *traditional global* or local methods.

Originally, DIALIGN has been developed as an all-purpose *multiple-alignment* program, but it soon became clear that it is particularly useful to align genomic sequences. In fact, it was the first program that was able to align genomic DNA sequences where islands of local sequence *conservation* are located between *nonrelated* sections of the sequences. DIALIGN has been used in many studies in *comparative* genomics.

In a *pioneering* series of papers, Gottgens et al. used the program to identify small *regulatory* elements by comparing upstream regions from various *vertebrate* genomes. Other applications of DIALIGN to comparative genomics include identification of sequence signatures for *pathogen* detection and gene finding in *eukaryotes.* A *systematic* evaluation and *comparison* of software programs for multiple alignment of genomic sequences has been carried out by Pollard et al..

SELECTED FEATURES OF THE PROGRAMME

Resemblance at Different Levels

DIALIGN has a number of program features that are useful for *aligning* genomic sequences. The *original version* of the program had two different options to compare DNA sequences, a *nucleotide*-level option, where (local) sequence *similarity* is measured by *comparing* sequences in a base-by-base fashion and a peptide-level option, where (local) segments of DNA sequences are translated into peptide segments according to the *genetic code,* and the similarity of the DNA segments is then calculated based on the *similarity* of the implied peptide segments.

This second option is useful if *distantly* related *protein-coding* sequences are aligned where sequence similarity may have been eroded at the *nucleotide* level but is still detectable among the corresponding proteins. Note that, in principle, the program compares every possible fragment alignment, i.e., every possible pair of equal-length sequence segments, thereby checking for sequence similarity in all possible reading frames. Because the program tries to identify a set of fragments with *maximum* total score, it selects those segment pairs that have the strongest degree of similarity.

This way, the program output implicitly contains some information about the likely reading frame in *protein-coding* sequences. In the *original program* version, the user had to decide if nucleic-acid sequences were to be compared at the *nucleotide* level or at the peptide level. We later implemented a so-called mixed-alignment option, where local sequence similarity is measured automatically at three different levels, namely (1) at the nucleotide level, (2) at the peptide level on the forward strand, and (3) at the peptide level on the reverse complement; the score of a fragment is then defined by the level of sequence *similarity* is (*locally*) *strongest*.

As a result, some regions in the *resulting* alignment may be exhibit stronger similarity at the peptide level in one of the two possible orientations, whereas other regions are more *strongly* related at the nucleotide level. The *mixed-alignment* option, together with the information about the potential reading frame contained in the output alignment, is useful for *gene-prediction* purposes; we recently used this option to obtain additional *extrinsic* information for the HMM-based gene finder AUGUSTUS.

Applicable Alignment

Practically all software tools for sequence alignment are fully automated. This means that the user enters the input sequences and can adjust a number of program parameters, but there is no direct way of influencing the *resulting* alignment. However, there are many reasons why such *automated alignment* procedures may fail to produce biologically *meaningful* output *alignments*.

We developed an anchored alignment approach where the user can specify certain sites within the input sequences that are to be aligned in the output alignment. The remainder of the sequences are then aligned according to the *constraints* defined by these user-defined anchor points.

More specifically, the user can specify pairs of equal-length segments of the input sequences, i.e., gapfree local pairwise alignments that are to be contained in the output alignment. If it is not possible to integrate all these user-defined anchoring segment pairs into one output alignment, a suitable subset is *automatically* selected by the program.

To this end, a score is to be defined by the user for each proposed anchor point, in case of conflicting *anchor* points, the program prioritizes them according to the user-defined scores. The *anchored-alignment* option is useful to enforce *meaningful* alignments in situation where the program is not able to produce a meaningful alignment *automatically*, but some expert information about sequence *homologies* is available. Here, anchor points can help to use such expert information for limproved *alignment* quality. A second application of our anchoring option is to speed-up the alignment procedure for long genomic sequences. Because the program has *originally* been designed to *align* rather small input sequences, its standard version if too slow to do *large-scale* alignment of genomic sequences.

Alignment of long sequences, however, can be accelerated considerably if known or *presumed homologies* are used as anchor points to *reduce* the alignment search space. To speed-up *alignment* of genomic sequences, a fast ocal *alignment* search tool such as BLAST or CHAOS can be used to identify anchor points.

RESULTS OF INPUT AND OUTPUT PROGRAMMES

DIALING and CHAOS at GOBICS

At Gottingen Bioinformatics Compute Server (GOBICS), we installed a website interface to use DIALIGN for multiple alignment of *genomic* sequences. At our website server, the previously explained anchoring option is used to speed-up the alignment *procedure*; anchor points are *identified* using the local alignment tool CHAOS. Alignments are *visualized* using the visualization tool ABC developed by Cooper et al..

This tool gives a *graphical* overview about the produced alignment and allows the user to *interactively* zoom in to inspect the alignment in detail. In addition, the produced alignment is returned in various formats. An online user guide gives all necessary information about input options and output format.

Command-Line Version of the Program

The source code of DIALIGN is freely available under the GNU *open-source* licence agreement. If the program is locally installed, more options are available than on the web server. The following program options are available for the *command-line* version; the corresponding program *parameters* are *explained* in the user guide that comes with the program.

Options for the Program Run

1. For DNA sequences, (local) sequence similarity can be calculated at the *nucleotide* level, at the peptide level and at both levels with the *"mixed-alignment"* option as *previously* explained. If the *"peptide-level"* or *"mixed-alignment"* option is used, it is possible to translate sequence segments only at the forward strand or to have the *program* look at both, the forward strand and the reverse *complement*.
2. There is a *threshold* parameter T that can be used to filter out all local sequence *similarity* with score below T.
3. Various options are available to further speed-up the program, but possibly at the *expense* of alignment sensitivity. For example, it is possible to reduce the *maximum* possible length of the fragments that the program uses to compose output *alignments*.
4. Anchor points can be specified using a command-line parameter and a special file containing coordinates and scores for the proposed anchor points. Details are explained on our webpage or in the user guide that comes with the *downloadable* program version.
5. As previously explained, the basic building blocks of DIALIGN are local gapfree pairwise alignments, i.e., local sequence similarity is primarily measured between pairs of sequences, and *homologies* involving more than two sequences are detected only as a result of such pairwise comparisons. For multiple alignment, there is a program option that prefers local similarities occurring in more than two sequences above those similarities that exist only between two input sequences.

This so-called overlap-weight option generally improves alignment quality, but it is *computationally* expensive. By default, this option is therefore used, if not more than 35 sequences are aligned; for larger data sets, this option is switched off. With two special program options, the user can use or switch off overlap weights independent of the input data size.

Alternative Alignments

1. In additional to the human-readable DIALIGN alignment format, the program can return the output alignment in FASTA or MSF format.
2. The program can return a list of all fragments (i.e., local pairwise gap-free alignments) that have been used to construct the output alignment. This list *contains* useful information about the fragments, e.g., their coordinates and similarity scores.
3. For multiple alignment, it is possible to output a list of all fragments contained in the respective pairwise alignments, no matter if they have been finally selected for the multiple alignment (*consistent fragments*) or not (*nonconsistent fragments*). This output file distinguishes between *consistent* and *nonconsistent fragments.*
4. It is possible to output all fragments (including *coordinates, fragment scores,* and so on) that have been considered by the program during the *alignment* procedure, though this list can be very long.
5. In the standard program output, *nucleotides* that remain unaligned, i.e., nucleotides not involved in any of the fragments selected by the program, are printed in lower-case letters, whereas aligned residues are printed in upper-case letters. It is possible to mask *unaligned* residues by using a special character instead of printing them in lower-case letters.
6. Two different output options are available to indicate the degree of (relative) local sequence similarity in the *alignment.*

Index

A

B

C

D

E

F

H

I

J

K

L

M

N

O

P

Q

R

S

T

U

V

W

X

Y

Z